Environmental Toxicology III

THIRD INTERNATIONAL CONFERENCE ON
ENVIRONMENTAL TOXICOLOGY

Environmental Toxicology III

CONFERENCE CHAIRMEN

V. Popov
Wessex Institute of Technology, UK

&

C.A. Brebbia
Wessex Institute of Technology, UK

INTERNATIONAL SCIENTIFIC ADVISORY COMMITTEE

Organised by
Wessex Institute of Technology, UK

Sponsored by
WIT Transactions on Ecology and the Environment
The International Journal of Sustainable Development and Planning

WIT Transactions

Transactions Editor

Carlos Brebbia
Wessex Institute of Technology
Ashurst Lodge, Ashurst
Southampton SO40 7AA, UK
Email: carlos@wessex.ac.uk

Editorial Board

Environmental Toxicology III

Editors

V. Popov
Wessex Institute of Technology, UK

&

C.A. Brebbia
Wessex Institute of Technology, UK

WIT*PRESS* Southampton, Boston

V. Popov
Wessex Institute of Technology, UK

C.A. Brebbia
Wessex Institute of Technology, UK

Published by

WIT Press
Ashurst Lodge, Ashurst, Southampton, SO40 7AA, UK
Tel: 44 (0) 238 029 3223; Fax: 44 (0) 238 029 2853
E-Mail: witpress@witpress.com
http://www.witpress.com

For USA, Canada and Mexico

Computational Mechanics Inc
25 Bridge Street, Billerica, MA 01821, USA
Tel: 978 667 5841; Fax: 978 667 7582
E-Mail: infousa@witpress.com
http://www.witpress.com

British Library Cataloguing-in-Publication Data

A Catalogue record for this book is available
from the British Library

ISBN: 978-1-84564-438-3
ISSN: 1746-448X (print)
ISSN: 1743-3541 (online)

The texts of the papers in this volume were set individually by the authors or under their supervision. Only minor corrections to the text may have been carried out by the publisher.

No responsibility is assumed by the Publisher, the Editors and Authors for any injury and/or damage to persons or property as a matter of products liability, negligence or otherwise, or from any use or operation of any methods, products, instructions or ideas contained in the material herein. The Publisher does not necessarily endorse the ideas held, or views expressed by the Editors or Authors of the material contained in its publications.

Printed in Great Britain by Martins the Printers.

Preface

The efforts of modern societies to design and manufacture goods that make human lives easier and more comfortable have produced a story of success in terms of offering constant improvements to our lifestyle. There is another side to this story, and that is the one which reports increased risks to human health and environment due to presence of various harmful substances in the environment. It is questionable whether there is any part of the environment that has been left unchanged and unaffected due to these activities. In many cases close to the human settlements the risks to health increase due to various stressors. Yet, many of us have to accept living cities with the commodities that they offer and the health risks that they create.

Environmental toxicology is an interdisciplinary science which integrates biology, microbiology, chemistry, engineering, environmental sciences, ecology and other sciences. Assessment of the environmental effects of chemicals is complicated as it depends on the organisms tested and involves not only the toxicity of individual chemicals, but also their interactive effects, genotoxicity, mutagenicity and immunotoxicity testing. Various stressors affect the environment sometimes showing synergistic effects which are very difficult to quantify or predict. These threats require more experimental and theoretical developments in order to produce approaches for characterization and appropriate strategies and assays for screening in order to detect the harmful agents and prevent them from reaching sensitive endpoints.

The Environmental Toxicology Conference created an atmosphere which encouraged fruitful interactions and exchange of knowledge and ideas amongst the participants working in industry and government and those employed at universities and research organizations.

This volume contains the edited contributions presented at the third Conference on Environmental Toxicology, which was held in Cyprus in 2010. The conference was organized by the Wessex Institute of Technology. It was sponsored by WIT

Transactions on Ecology and the Environment and The International Journal of Sustainable Development and Planning. The first Conference took place in Myconos Island, Greece in 2006 and the second one was held in Granada, Spain in 2008.

The editors would like to thank all the authors for their papers and, in particular, the members of the International Scientific Advisory Committee for their help during the review process.

The Editors,
Cyprus, 2010

Contents

Section 2: Ecosystem health

Section 3: Biodegradation, bioremediation and biomonitoring

Section 4: New trends in environmental toxicology

Section 1
Environmental health risk

Toxicity of volatile organic compounds (VOCs) mixtures using human derived cells

C. Khalil & J. Nasir
School of Safety Sciences,
Chemical Safety and Applied Toxicology Laboratories, UNSW, Australia

Abstract

Assessing the effects of contaminants is an issue of high priority for governmental safety health and environmental agencies around the world. The general conservative consensus is that chemicals in mixtures interact by concentration addition. However, previous studies also report that concentration addition of mixture components does not always reflect the overall toxicity of a mixture. Volatile organic compounds (VOCs) such as Benzene, Toluene, Xylene and Formaldehyde (BTXF) belong to the air pollutants found in urban and indoor environments. They could trigger acute and chronic adverse health effects like allergy, respiratory and cardiovascular diseases. The volatile nature of these compounds poses additional problems in assessing individual volatile chemical toxicity let alone mixtures of these chemicals. Our research aims at establishing the true toxic effects of VOC exposure *in vitro* using a static direct exposure glass-chamber method. This was achieved by assessing and comparing individual and interactive effects of VOCs in exposed human epithelial lung (A549) and liver cells (HepG2) using the MTS cytotoxicity assay to assess cell viability upon VOC insult. The study results clearly indicated the limitation of the concentration addition method used in assessing volatile mixtures cytotoxicity and the need to develop new techniques for rapid and accurate mixture toxicity determination. The study may have implications for regulatory risk assessment of environmental volatile organic chemicals.
Keywords: static method, MTS, cytotoxicity, lung cells, liver cells, VOCs.

1 Introduction

Human environmental chemical exposures are characterised by exposures to direct multiple chemical combinations or sequential exposure to individual or

WIT Transactions on Ecology and the Environment, Vol 132, © 2010 WIT Press
www.witpress.com, ISSN 1743-3541 (on-line)
doi:10.2495/ETOX100011

different chemicals at low concentrations [1]. These exposures are unavoidable because chemicals represent an integral part of our life and play an important role in promoting human lifestyle and wellbeing. However, data on chemical mixtures is sparse due to the focus of traditional toxicology on individual chemicals and their toxicities, but interest in this emerging area of mixture toxicity has been building in recent years [2].

The main objective of this research is to present findings relevant to assessment of toxicological human health risks as a result of exposure to volatile mixtures. This is achieved through toxicity determination of individual chemical components in a mixture and correlating that to the observed toxicity resulting from the mixture. The selection of biological tests rather than chemical identification as a starting point is based on the Australian and New Zealand Environment and Conservation Council (ANZECC) environmental guidelines for conservation and sustainable development which recognises a hierarchy of evidence based assessment. This assessment ranges from most powerful (biological effects) to least powerful (chemical identification and measurement) evidence [3].

Furthermore, toxicity assessment of chemical mixtures is a challenging task and requires understanding interactions and characteristics of the chemicals present [4]. Unfortunately in the environment, it is not always possible to identify all individual chemicals in a mixture and their interactions [5], hence the need for a rapid, repeatable and accurate *in vitro* screening technique (e.g. static glass-chamber method) to assess overall toxicity of mixtures. Natural systems are complex systems and it is virtually impossible to understand the full mechanism of migration, accumulation, biotransformation and toxicity of volatile chemicals introduced in such open systems, but it is hoped that the results presented here could elucidate some of the toxic potential of mixtures in a controlled environment [5].

This study investigated airborne concentrations of Benzene, Toluene, Xylene and Formaldehyde and their mixtures. The toxicity assessment was undertaken in human derived cells (lung and liver cells) using a colorimetric assay, the MTS assay. The investigation endpoint aimed at producing dose response curves (for individual and chemical mixtures) to establish the nature of the toxicological effects resulting from chemicals mixtures exposure. The cell culture selection was based on potential targets of exposure in humans (mainly lung and liver). The selected cells by virtue of their location, numbers and ease of growth in culture could be used as possible indicators of cellular damage caused by multiple contaminant exposure *in vitro* [6]. Furthermore, the techniques used are rapid, reproducible and generate accurate individual and mixtures toxicity profiles within hours of conducting a full set of assays.

2 Materials and methods

2.1 Chemicals

Benzene (CAS No. 71-43-2) was purchased from BDH Chemicals, Australia (Laboratory reagent). Toluene (C6H5CH3), CAS# 108-88-3, was purchased from APS Finechem, Australia (Analytical reagent). Xylene (CAS No. 1330-20-7) and Formaldehyde (CAS No. 50-00-0) both were purchased from Chem-Supply Australia (Laboratory reagents). *In vitro* assay reagents were purchased from Promega (USA) and Sigma (USA).

2.2 Human-derived cell cultures

The main cell cultures used in this research consisted of an epithelial lung carcinoma cell (A549) [ATCC CCL-185] and a Heptacarcinoma human cell line (HepG2) derived from the tissue of a 15 year old Caucasian male [ATCC HB8065]. Cells were sub-cultured as adherent cells in 75 cm^2 tissue culture flasks with 0.2 μm vented seals (Falcon). The culture media consisted of colour free Dulbeco's modified eagle medium (DMEM): RPMI 1640 (1:1) purchased from Sigma Chemicals, supplemented with 5% foetal calf serum (Trace Bioscience), 3% Sigma antibiotics [penicillin (100 U/ml), streptomycin (0.1 mg/ml) and L-glutamine (2 mM)]. Cell lines (HepG2 and lung cells (A549) were cultured at 37°C at sub-confluence in a humidified incubator set to a mixture of 5% CO_2/95% air. Cell viability was over 95% as measured by tryptan blue dye exclusion. Confluent cells in log phase of growth were released from the bottom of the culture flask using Trypsin EDTA (Gibco, USA), and then washed three times with cell culture medium before being seeded on porous membranes (0.4 mm) on snapwell inserts.

Snapwell insert is a modified transwell culture insert with a 12 mm diameter providing a growth area of 1.12 cm^2 (clear polyster Snapwellt insert, 3801, Corning), supported by a detachable ring that was placed in a six well culture plate. Culture media and 1% (v/v) HEPES buffer was added to both sides (bottom, 2 ml; top, 0.5 ml) of the membranes. The snapwell inserts in six well plates were incubated at 37°C for one hour as an initial equilibrium time to improve cell attachment. Culture media was then removed from the top and replaced with fresh culture media (0.5 ml) containing a cell suspension, (20–30) x 10^4 cells, supplemented with 5% FCS, 1% antibiotics and 1% HEPES buffer. Cell cultures in six well plates were incubated at 37°C in a humidified incubator for 24 h. Cell attachment was observed under the light microscope (Leitz Wtzlar, Germany), medium was removed from both sides of the snapwell inserts and membranes washed with Hank's balanced salt solution (HBSS; Gibco, USA) from both sides (top, 0.5 ml; bottom, 2.0 ml). Cells on the membranes were exposed to airborne concentrations of test chemicals on their apical side while being nourished from their basolateral side, using the static exposure technique.

2.3 Exposure protocol

Standard test atmospheres were generated using a static methodology as outlined in Bakand et al. [7]. Briefly, a known quantity of volatile liquid was introduced into the glass bottle onto a filter paper. Human derived cells grown on snapwell inserts were detached from their holders and placed into sterile individual glass wells. Each glass well contained 1.2 ml of serum free culture media supplemented with 1% HEPES buffer. Aliquots of test chemicals (ranging from 0, 2.5, 5.0, 10.0, 15.0, 20.0, or 30.0 ml) were introduced to the glass chambers onto the filter paper. Glass chambers were immediately closed, sealed with parafilm and placed on an orbital mixer incubator (50 RPM; Ratek Instruments, Australia) at 37°C. Each aliquot of volatile liquid was introduced into a single chamber. Human cells were exposed to various airborne concentrations of volatile test chemicals directly at the air/ liquid interface for 1 h. Details on airborne test concentration calculations can be found in [8].

At the end of the exposure time, snapwell inserts were removed and replaced in their holders within six well plates, Culture media supplemented with 1% HEPES buffer was added to both sides (top, 0.5 ml; bottom, 2 ml) of the membranes. Cells were incubated for 24 h at 37 °C in a humidified incubator. At the end of the incubation time, cell viability was investigated using the MTS (Tetrazolium salt) assay.

For each *in vitro* experiment, two controls were set up in identical conditions including an IC_0 (0% inhibitory concentration; cells only) and an IC_{100} (100% inhibitory concentration; media only), and exposed to air only during the exposure time.

2.4 Toxicity determination of individual chemicals and their mixtures

Test chemicals were freshly prepared each time, immediately before use. Range finding experiments were undertaken to determine the concentration range of individual test chemicals that inhibited the viability of 50% exposed cells. In general for any binary, ternary, quinary mixtures of selected toxicants, the previously experimentally derived IC_{50} of each toxicant in the mixture was used as an index of toxicity. The toxicants were prepared such that their fractional effects individually in the mixture was calculated in proportion of their concentration to the total concentrations of all toxicants in the mixture and such that the sum of all ratio combinations equalled to a theoretical additive value of 1. E.g. for chemicals A, B and C;

$$\frac{IC_{50}(A)}{IC_{50}(mix)} + \frac{IC_{50}(B)}{IC_{50}(mix)} + \frac{IC_{50}(C)}{IC_{50}(mix)} = 1 \tag{1}$$

IC_{50} (A, B or C) denotes 50% inhibitory concentration for chemical A, B or C when administered individually and calculated with the respective *in vitro* tests.

$IC_{50\ (mix)}$ is the sum of the total individual IC_{50} concentrations for the chemicals. For chemicals A, B and C, for e.g. the mixture ratio of chemical

$$A = 1 - (IC_{50B}/IC_{50mix} - IC_{50C}/IC_{50\ mix}) \tag{2}$$

and so forth the remaining chemicals. After having the IC_{50} values of individual chemicals, ratios of mixture components were calculated using the equation above and then IC_{50} values were determined for binary, ternary and quinary mixtures as per procedure described above and dose response curves were generated accordingly (Fig 1 & 2). Once IC_{50} values of individual and chemical mixtures were calculated, interaction effect was estimated using isobole method briefly discussed in a later section.

2.5 Cytotoxicity assays

The MTS assay (Promega, USA) was selected for measuring the number of active cells in the culture (based on the lactate dehydrogenase activity in the mitochondria). The MTS assay measuring the conversion of a soluble tetrazolium salt to a formazan product by viable cells [9]. The assay consisted of an MTS solution prepared by mixing a solution of MTS (42 mg MTS powder in 21 ml of DPBS pH 6.0-6.5) with a PMS solution (0.92 mg/ml PMS in DPBS) to the cells to be tested in a ratio of 1:5. The MTS was then incubated with the cells for a period of 2 h at 37°C in the dark. After 2 h, the cellular supernatant absorbance was measured. The amount of reduced Formazan was assessed by measuring the optical density at 492 nm using a Labsystem Multiskan MS plate reader. Data was plotted as a dose response curve exposure versus absorbance reading

2.6 Statistical analysis

Dose response curves reported were plotted from experimental data. All data reported was expressed as mean ± SD of 3-4 replicated wells. Statistical procedures and graphical analysis were performed using Graphpad Prism software.

2.7 Results

The dose response curves for individual VOCs and their mixtures were generated using the static methodology as outlined in Figures 1 & 2. The graphs are mainly presenting the experimental data from a series of experiments using airborne concentrations of individual and mixtures of generated VOCs.

Once IC_{50} values of individual and chemical mixtures were calculated, interaction effect was estimated by using isobole method [10]. The isobole

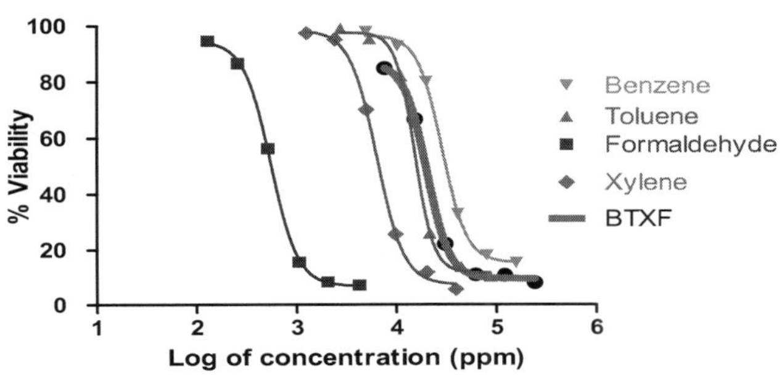

Figure 1: Individual and mixtures cytotoxicity in lung cells (A549).

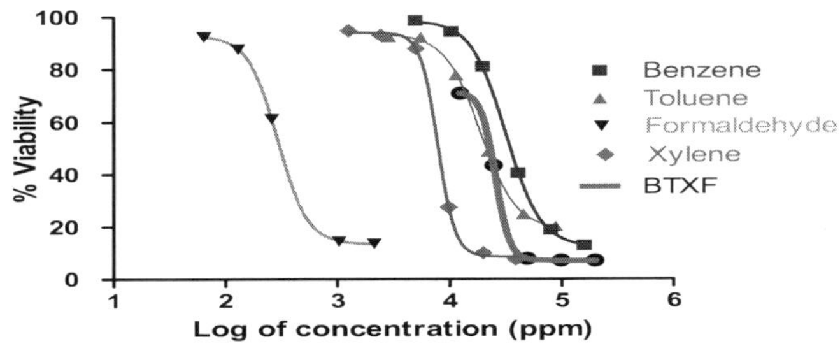

Figure 2: Individual and mixtures cytotoxicity in liver cells (HepG2).

method is based on the assumption that if A and B are applied jointly their mixture toxicity can be estimated by dividing the concentration of each toxicant in the mixture with the concentration of the toxicants applied singly that yields the same effect as the mixture. The method is valid for mixtures of any given number of toxicants [11]. Mathematically, isobole method for an additive mixture effect can be described as:

$$\frac{dA}{DA} + \frac{dB}{DB} = 1 \tag{3}$$

dA and dB = the dose of chemical A and B in the mixture which produces a given effect while DA and DB = the dose of chemical A and B in single toxicant experiments which elicits the same effect as the mixture.

If the isobole calculation yields a figure less than 1 then the relationship is synergistic. Furthermore if the calculation is more than 1 then the relationship can be classified as antagonistic (Table 2).

Table 1: Chemical toxicity parameters.

Chemicals	**Lung Cells (A549)** IC_{50} (ppm)	**Liver Cells (HepG2)** IC_{50} (ppm)
Benzene	29915±1103	33113±1250
Toluene	14217±1132	17721±126
Xylene	6847±792	7453±830
Formaldehyde	524±105	305±84
Benzene: Toluene	29509±563	4634±1198
Benzene:Xylene	29006±849	28183±1102
Toluene: Xylene	18465±746	20550±716
Benzene: Formaldehyde	17715±208	14043±388
Benzene:Toluene:Xylene	34062±2626	33424±1988
Benz:Tol:Xyl:Form	21999±1775	28739±2957

Table 2: Interactive effects of VOC mixtures.

Chemical Interaction	*A 549 Cells*		*Hep G2 Cells*	
	Ratio	**Interaction Effect**	**Ratio**	**Interaction Effect**
Benz:Toluene	1.31 –1.35	Antagonistic	1.71 - 1.79	Antagonistic
Benz:Xylene	1.56 - 1.58	Antagonistic	1.32 - 1.48	Antagonistic
Toluene:Xylene	1.74 - 1.76	Antagonistic	1.60 – 1.68	Antagonistic
Benz:Formaldehyde	0.90 - 0.92	Synergistic	0.82 - 0.92	Synergistic
Benz:Tol:Xylene	1.85 - 2.16	Antagonistic	1.61 – 1.86	Antagonistic
Benz:Tol:Xylene: Formaldehyde	1.54 – 1.80	Antagonistic	2.2 – 2.5	Antagonistic

3 Discussion

There are a large number of known chemicals in the natural and built environment that humans are exposed to in addition to an ever increasing number of new chemicals and mixtures for which no data exists [12]. The lungs, skin, central nervous system (CNS), liver and the kidneys are the main body systems affected by these chemicals [13]. Testing, even for the most potent mixtures with classical toxicological protocols, is unrealistic and perhaps not achievable. Most federal agencies and international organisations such as ATSDR, US EPA, NIOSH use a default assumption of response additivity in assessing mixture toxicity in exposed human populations [10, 14]. However this assumption has setbacks as it does not factor chemical interactions in toxicity determination. There is also a lack in direct comparison between the available approaches for mixtures toxicity assessment, as most tend to differ in data types and nature of observations as reported in the literature [15]. No published *in vitro* airborne toxicity data could be sourced for mixtures of benzene, xylene, toluene and formaldehyde. However, inhalational *in vivo* toxicity data for Toluene and Xylene have been reported in rat by the NIOSH Registry of Toxic Effects of Chemical Substances (RTECS).

Figures 1 and 2 indicated the toxicity of individual VOCs and their mixtures. The striking feature in the two graphs is the non additive effect of BTXF mixtures (when compared to individual VOC curves) which suggested a non additive effect resulting from ternary mixtures. The results of individual chemicals (Table 1) reported the cytotoxicity of formaldehyde in liver cells (IC_{50}=305 + 84 ppm) was quite higher than that in lung cells (IC_{50} = 524 + 105 ppm). Among structurally similar VOC's, the cytotoxicity of xylene (IC_{50}= 6847± 792 ppm, A549 cells; 7453±830 ppm, Hep-G2) was found 2-3 times higher than that of toluene (IC_{50} = 14216± 132 ppm, A549; 17721±1226 ppm, HepG2) and five times higher than benzene (IC_{50} = 29915±1103 ppm, A549; 33113±1250 ppm, HepG2). Among four selected chemicals, low weight carbonyl compound, formaldehyde was more toxic while cytotoxic effect of well-recognised carcinogenic compound i.e. benzene was less than other chemicals. The LC_{50} (50% Lethal Concentration) values of xylene (5000 ppm) and toluene (13,000 ppm) have been reported in rats following 4 h exposure [16, 17]. Based on *in vitro* results, IC_{50} values for xylene (6847±792 ppm) and toluene (14217±1132 ppm) for human lung cells were determined after 1 h exposure (Table 1). An *in vitro/in vivo* comparison indicates that the *in vitro* toxicity findings in the present study are in good correlation with inhalational *in vivo* published data for both volatile organic solvents. The LC_{50} for benzene (10,000 ppm) have been reported in rats following 7 h exposure [18]. The observed *in vitro* IC_{50} values for benzene (1 h exposure) are 29915±1103 ppm (A549 cells) and 33113±1250 ppm (HepG2 cells) which clearly indicates the higher sensitivity achieved by using the developed *in vitro* methodology. Regarding formaldehyde, the lethal concentration (LC_{50}) for inhalation s reported in the literature ranged from 405 ppm (497 mg/m^3) in mice to 471 ppm (578 mg/m^3) in rats following four hours exposure [19]. In comparison, in this study IC_{50} values for *in vitro* toxicity of formaldehyde in A549 cells and HepG2 cells were found to be 524±105 ppm and 305±84 ppm respectively. *In vivo* published data suggest that the cytotoxic effects of formaldehyde in laboratory animals appears to be more closely related to the exposure level of formaldehyde than to the time of exposure or total dose [20, 21].

Binary and ternary mixtures of benzene, toluene and xylene showed antagonistic effects on human lung and liver cells. A study by Ewa and Anna (2008), on binary effect of toluene and xylene on lipid peroxidation also reported an antagonistic effect [22]. Briefly, the binary mixtures of benzene: formaldehyde exerts synergistic effects in human lung and liver cells while binary and ternary mixtures of structurally similar VOCs *i.e.* Benzene: Toluene: Xylene demonstrated antagonistic effects on both types of human cells (Table 2). Toxicity of quinary mixtures of BTXF was higher (A549 IC_{50}: 21999±1775; HepG2 IC_{50}: 28739± 2957) when compared to ternary mixtures of BTX (A549 IC_{50}: 34062±2626; HepG2 IC_{50}: 33424±1988). This may be due to presence of formaldehyde in quinary mixture however combined effect of BTX was dominant over formaldehyde toxicity hence overall effect of BTXF quinary mixture was antagonistic.

It is concluded from the data presented there are possible toxicological interactions (i.e. departures from additivity) that have clear implications for risk assessment. The presented data clearly highlighted the limitations of an additive interaction assumption and the need to focus on volatile chemical mixture studies. This is an important factor to consider because real life exposures consist of exposure to a cocktail of numerous chemicals rather than single individual chemical. The *in vitro* cytoxicity studies conducted are relevant and important to risk assessment of chemical mixtures in several ways. The results showed that chemicals in a mixture do not necessarily act in an additive fashion and the possible inclusion of cytotoxicity assays can help in the regulatory decision making process. The study may have implications for risk assessment of environmental exposure and establishing safe levels of exposure. Studies are being conducted looking at a dynamic exposure to these VOC mixtures and their immunotoxic/genotoxic effect on human health.

Acknowledgement

The project was funded by the Faculty of Science research grant initiative, UNSW.

References

[1] Boekelheide, K. (2007). Mixed messages. Toxicol. Sci. 99, 1–2.
[2] Feron, V. J., and Groten, J. P. (2002). Toxicological evaluation of chemical mixtures. Food Chem. Toxicol. 40, 825–839.
[3] ANZECC. Australian and New Zealand guidelines for the assessment and management of contaminated sites. Australian and New Zealand Environment and Conservation Council / National Health and Medical Council, 1992.
[4] Azzi, R., Hayes, A., Khalil, C., & Winder, C. An in vitro study of the interactive effect of 24 binary and ternary mixtures from the GHS classification groups. ALTEX 22 (Spl):128, 2005.
[5] Belden, J.B., Gilliom, R.J., Martin, J.D., Lydy, M.J. Relative toxicity and occurrence patterns of pesticide mixtures in streams draining agricultural watersheds dominated by corn and soybean production. Integrated Environmental Assessment & Management. 3(1):90-100, 2007.
[6] Khalil, C. Combining Three *in vitro* assays for detecting early signs of UVB cytotoxicity in cultured human skin fibroblasts. WIT Transactions on Biomedicine and Health (10): 349-359, 2006.
[7] Bakand, S., Winder, C., Khalil, C., Hayes, A novel in vitro exposure technique for toxicity testing of selected volatile organic compounds. J. Enviro. Monitoring. (8): 100-105.
[8] Salvatore R. DiNardi, 2003, Occupational Environment: Its Evaluation, Control, And Management, 2nd Edition, ISBN-13: 978-1-931504-43-1
[9] Promega. CellTiter 96® Aqueous Non-Radioactive Cell Proliferation Assay: Technical Bulletin No 169. Promega Corporation. Madison, USA.
[10] ATSDR (2004). Guidance manual for the assessment of joint toxic action of chemical mixtures. Agency for Toxic Substances and Disease Registry. US Department of Health and Human Services, Atlanta].
[11] Poch, G., R. J. Reiffenstein and H. D. Unkelbach (1990). "Application of the Isobologram Technique for the Analysis of Combined Effects with Respect to Additivity as Well as Independence." Canadian Journal of Physiology and Pharmacology 68. pp 682-688
[12] Marinovich, M., Ghilardi F. and Galli C. L. (1996). "Effect of Pesticide Mixtures on in Vitro Nervous Cells: Comparison with Single Pesticides." Toxicology 108: 201-206.
[13] Winder, C., 2004a. Occupational respiratory diseases. In: Occupational Toxicology, Second edition. Winder, C. and Stacey, N. H. (Eds). CRC Press, Boca Raton, pp. 71-114.
[14] U.S. EPA. (2000) Supplementary guidance for conducting health risk assessment of chemical mixtures. EPA/630/R-00/002

[15] Pounds, K. G., Haider J., Chen D. G. and Mumtaz M. (2004). "Interactive Toxicity of Simple Chemical Mixtures of Cadmium, Mercury, Methylmercury and Trimethyltin: Model-Dependent Responses." Environmental Toxicology and Pharmacology. 18: 101- 113.
[16] NIOSH, The Registry of Toxic Effects of Chemical Substances, Toluene, RTECS # XS5250000, 2004.
[17] NIOSH, 2004b, The Registry of Toxic Effects of Chemical Substances, Xylene, RTECS # ZE2100000, 2004.
[18] ATSDR (2007), Toxicological Profile for Benzene. U.S. Department of Health and Human Services: Agency for Toxic Substances and Disease Registry, Atlanta, GA.
[19] IPCS, 1989. Formaldehyde. Environmental Health Criteria No 89. International Programme on Chemical Safety. World Health Organization, Geneva.
[20] IARC, 1995. International Agency for Research on cancer- Summaries & Evaluations, Formaldehyde, Vol 62.
[21] Wilmer, J. W., Woutersen, R. A., Appelman, L. M., Leeman, W. R. and Feron, V. J., 1989. Subchronic (13-week) inhalation toxicity study of formaldehyde in male rats: 8-hour intermittent versus 8-hour continuous exposure. Toxicology Letters, 47:3, 287–293.
[22] Ewa Sawicka and Anna Długosz, Toluene And P-Xylene Mixture Exerts Antagonistic Effect On Lipid Peroxidation In Vitro, International Journal of Occupational Medicine and Environmental Health 2008; 21(3):201–209.

Carcinogenesis in female C57Bl/6J mice chronically exposed to sodium arsenate (As^V) in drinking water for 2 years

M. Krishnamohan[1], A. A. Seawright[1], M. R. Moore[1,2] & J. C. Ng[1]
[1]*The University of Queensland, National Research Centre for Environmental Toxicology (Entox), Australia*
[2]*Water Quality Research Australia, Adelaide, Australia*

Abstract

Arsenic is a ubiquitous element in the environment and has been classified as a human carcinogen primarily based on epidemiological evidence. It has been estimated there are over 100 million people globally being exposed to elevated arsenic from both natural and anthropogenic sources. Surprisingly, positive carcinogenicity animal studies were lacking until recent years. We aim to validate inorganic arsenate carcinogenic effect in C57Bl/6J mice, and establish the dose-response relationship using environmental concentrations of arsenic similar to those found in typical endemic-areas. Mice were given 0, 100, 250 or 500 µg As/L in the form of sodium arsenate in drinking water *ad libitum* over 2 years. Tumours occurred after about 18 months of arsenic exposure otherwise the animals appeared to be normal in their appearance and behaviour. Incidences of all types of tumours and non-tumourous lesions in the treated groups were higher than those observed in the control group. The induction of tumours was in a dose-response manner for some tumour types. Enlargement of the mesenteric lymph node due to hyperplasia or neoplasia of lymphoid elements was commonly observed. Apart from abdominal cavity lymph nodes, tumours were frequently observed in the liver, spleen and intestinal wall, and to a lesser extent in the lung with various other tissues also occasionally affected. Of the non-tumourous lesions, haemorrhagic ovarian cysts occurred more frequently in the treated groups than in the control group. Our results suggest that the C57Bl/6J mouse model can be a useful adjunct for further mechanistic studies of arsenic carcinogenesis. This bioassay data may also be considered for the risk evaluation of chronic exposure to inorganic arsenic.

WIT Transactions on Ecology and the Environment, Vol 132, © 2010 WIT Press
www.witpress.com, ISSN 1743-3541 (on-line)
doi:10.2495/ETOX100021

Keywords: arsenic, arsenate, carcinogenesis, tumours, lymphoma, mice, drinking water, chronic exposure, risk assessment.

1 Introduction

Inorganic arsenic is classified as a human carcinogen [1, 2]. Over 100 million of people globally are at risk of exposure to elevated levels of arsenic in drinking water [3]. Although it is well known that arsenic is toxic to both humans and animals, the mechanism underlying its chronic toxicity remains unclear. Several long-term animal studies in which daily doses of sodium arsenate and sodium arsenite were given in the drinking water (up to 400 μg/L) had been found to be negative in rats, mice, beagles and cynomologus monkeys, suggesting the apparent non-carcinogenicity of arsenic in animal models [4].

Adenocarcinoma was induced in the stomach of the rats implanted with 8 mg of arsenic trioxide in a capsule, by surgical implantation [5]. Hamsters administered with 3 mg As/kg of arsenic trioxide using charcoal carbon and 2 mM H_2SO_4 (a carrier to increase retention) by intra-tracheal installation once weekly for 15 weeks had low incidences of carcinomas, adenomas, papillomas and adenomatoid lesions of the respiratory tract [6]. Though these earlier animal studies demonstrated the carcinogenicity of inorganic arsenic in animals, they could not be taken as reliable animal models [7].

However, in more recent years, there have been several positive observations showing the carcinogenic effects of inorganic arsenic in mice. The first inorganic arsenic carcinogenicity study with chronic, low-dose exposure was done by Ng et al. [8] in C57BL/6J mice exposed to 500 μg As/L sodium arsenate in drinking water for 2 years. The authors reported increased incidence of tumours in various organs of treated mice but not in the control group. In another study [9], sodium arsenate in the drinking water (0, 1, 10 and 100 mg/L) administered to male A/J mice for 18 months resulted in an increase of lung tumour multiplicity and size in a dose response manner. Sodium arsenite was also proved to a carcinogen via the transplacental pathway [10, 11]. In this current study we aimed to confirm the study by Ng et al. [8] and further evaluate the carcinogenic effect of inorganic arsenic by including a lower and wider range of sodium arsenate concentrations. The water arsenic concentrations are similar to those reported to have caused arsenicosis in As-endemic areas [12].

2 Materials and methods

Sodium arsenate (Na_2HAsO_4) was purchased from Ajax Chemical, Australia. Animal experimental protocols were approved by the Queensland Health Animal Ethics Committee (AEC No. NRC 2/99/19). Female C57Bl/6J mice, aged 4 weeks, were divided into six groups of 70 each, 5 mice per cage, and were given drinking water containing 100, 250 or 500 μg As/L as sodium arsenate *ad libitum* for 24 months. A group of 105 control mice was given demineralised water containing <0.1 μg As/L. Female mice were used in our previous inorganic arsenic carcinogenicity study [8]. The stock solution of sodium arsenate was prepared and kept in the refrigerator for one month. The working solutions of 100, 250 and 500 µg As/L were prepared every week. Five animals from the control group were sacrificed at time zero, while 5 animals from the control and all the treatment groups were sacrificed after 2, 6 and 12 months exposure for the biomarker studies [13]. The animal care facility was operated at a controlled temperature set at 21–23 °C, with 13 filtered air changes per hour, a 12/12 h light/dark cycle, and year-round relative humidity of approximately 60%. All animals were kept in standard polycarbonate cages with stainless steel wire-mesh tops equipped with polycarbonate plastic drinking bottles and stainless steel sip-tubes, and given a commercial rodent diet *ad libitum* (Norco Pty Ltd, Brisbane, Australia).

Arsenic concentrations of the drinking water and rodent diet were monitored by HPLC–ICP–MS. The volume of the drinking water consumed by the mice in each cage, and the body weight of each mouse, were measured weekly. Five animals were sacrificed at time zero from the control group and five from each treatment group were sacrificed after 2, 6, 12 and 18 months for interim pathological examinations. Mice that became sick or which had developed significant skin lesions were also sacrificed by an overdose of carbon dioxide at various times up until the conclusion of the study at 104 weeks. Gross pathological changes, including tumours, were recorded and photographed. Tissue samples were collected for histopathological examination in a 10% buffered neutral formalin during necropsy. Paraffin sections were prepared from formalin-fixed tissues and stained with haemotoxylin and eosin.

3 Results

3.1 General observations

Tumours occurred after about 18 months of arsenic exposure; otherwise no abnormal appearance or behaviour was noticed in any of the animals. A total of 9 animals out of 85 were found dead in the control group; and 5, 8 and 11 animals out of 55 were found dead in the 100, 250 and 500 µg As/L groups respectively. No reason was found for the sudden death of some animals. Other animals died because of fatal haemorrhage from a blood-filled ovarian cyst or from large internal tumours.

3.2 Tumour incidence and gross pathology

Percentage incidences of all types of tumours and non-tumourous lesions in the treated groups were higher than those observed in the control group at the end of 2 years. Enlargement of the mesenteric lymph node due to hyperplasia or neoplasia of lymphoid elements was commonly observed. Apart from abdominal cavity lymph nodes, tumours were frequently observed affecting the liver, spleen and intestinal wall with various other tissues also occasionally affected. Of the non-tumourous lesions, haemorrhagic ovarian cysts were observed more frequently in the test groups than in the control group. Examples of tumours in

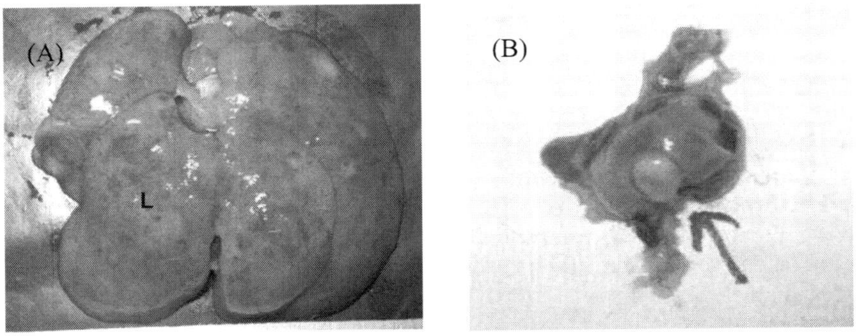

Figure 1: A mouse liver (A) diffusely invaded by tumour tissue. Pale tumour foci were surrounded by hyperaemic zones from the 500 µg As/L treatment group; a mouse lung (B) shows a large nodular tumour (arrow) from the 250 µg As/L treatment group.

Table 1: Percentage of tumour incidences observed in the control and sodium arsenate treated female C57Bl/6J mice at the end of 2 years.

Tumour incidence (%)	Control	100 µg/L	250 µg/L	500 µg/L
No. of mice exposed to As for more than 12 months	85	55	55	55
Mice with all types of tumours	15	25	62	56
Mice with lymphoma	11.7	15	31	29
Lymph nodes only	4.7	2	5.5	7
Organs & lymph nodes	7	13	25.5	22
Mice with other types of tumours	4	9	24	15
Mice with multiple tumours	1	2	7	13

the liver and lung are shown in Figure 1. There were significantly higher incidences of all types of tumours in the treated groups compared to the control group (Table 1) in a dose response relationship. Multiple tumours include lymphoma, plasmacytoma and histiocytic sarcoma.

Histological examination showed that lymphoma was the major type of tumour observed in both treated and control groups. Mesenteric lymph node was the major lymph node affected in the experiment animals. There was a significant difference in the incidence of lymphoma in the treated groups compared to the control group. A significant dose-response relationship was observed in total lymphoma incidence including lymph nodes and organs (p=0.0021), and lymphoma invading the organs (p = 0.0033). A significant dose-response relationship was observed in the incidences of other types of tumours (p = 0.0018) and multiple tumours (p = 0.0016).

After the mesenteric lymph node, liver and spleen were the major organs affected by As^V treatment. A higher incidence of liver, spleen and gastrointestinal tract lymphoma was observed in the treated groups compared to the control group. However, there was no evidence of a dose-response relationship (data not shown). Lung lymphoma was seen in all the treated groups but not in the control group. Uterus lymphoma was seen only in the 250 µg As/L group. Submandibular lymph nodes, thoracic lymph nodes, mammary glands, pancreas and subcutaneous lymph node were affected in at least one of the treated groups but not in the control group.

Plasmacytoma was observed in the liver, spleen, kidney, lung, GI tract, pancreas and subcutaneous glands of at least one of the treated groups but not in the control group. A higher incidence of histiocytic sarcoma was observed in the liver and spleen of treated groups compared to the control group. No urinary tract or bladder tumour was observed, except in one mouse in the 500 µg As/L group that developed bladder histiocytic sarcoma. Tubulosomal adenoma of the ovary was observed in the 250 and 500 µg As/L treated groups.

Other types of tumour such as Harderian gland adenoma, pituitary adenoma, subcutaneous fibrosarcoma and hepatocellular adenoma were found only in the treated groups and not in the control group. Hepatocellular carcinoma was higher in the 250 µg As/L group compared to the control group and it was not observed in the 100 and 500 µg As/L treated groups. Non-haematopoietic cancers observed are shown in Table 2.

3.3 Histology

Histological examinations (H&E) of the tumours showed various lymphoma patterns. These are illustrated in Figure 2.

Table 2: Percentage distribution of non-haematopoietic cancers in the control and sodium arsenate-treated mice.

Type	Control	100 µg/L	250 µg/L	500 µg/L
Hepatocellular carcinoma	1	0	4	0
Hepatocellular adenoma	0	2	9	5
Subcutaneous fibrosarcoma	0	0	2	0
Harderian gland adenoma	0	4	2	0
Pituitary adenoma	0	4	0	0

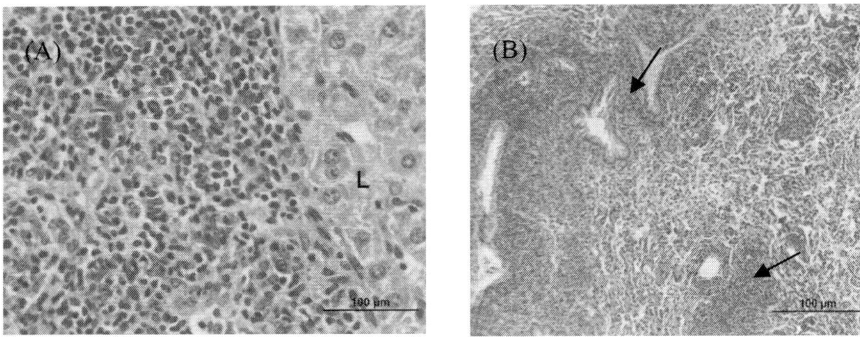

Figure 2: Lymphoid metastasis in the liver (A) from a mouse treated with 500 μg As/L. The cells are mostly small, dense and variable in shape. Low power illustration of lung (B) showing dense accumulations of lymphoid cells surrounding blood vessels and bronchi (arrows) from a mouse treated with 250 μg As/L as sodium arsenate in drinking water for two years.

4 Discussion

Although Dimethyl arsenic acid (DMA,) a major metabolite of inorganic arsenic, had been shown to cause bladder cancers in rats [14], inorganic arsenic alone has not been proved to be a carcinogen in any of the known animal models until recently [8]. The carcinogenic effect of arsenic has since been shown to be able to cross the placenta and subsequently induce tumours in the off-spring of pregnant dams exposed to relatively high concentrations of arsenic [10, 11]. Ng et al. [8] reported that when female C57Bl/6J mice were given 500 μg As/L as sodium arsenate (As^V) in the drinking water for up to 26 months, the tumour incidence in all organs was 41.1%. No tumours were reported in the control group. These findings were reported in the Environmental Health Criteria on Arsenic and Arsenic Compounds [15], which stated that "this was the first experimental carcinogenicity study in rodents using a relevant route of exposure and relevant exposure level" and that the incidence of tumours was "treatment related". In the report of this study, the types of tumours were not reported and no dose response effect was determined. The present study was designed to investigate whether the human carcinogen, inorganic arsenic is also carcinogenic in mice at concentrations which are commonly found in As-endemic areas and to confirm the previous study [8]. The overall incidence for all types tumours was 56% which is slightly higher but in general agreement with the tumour incidence (41.1%) reported previously [8]. More significantly our results in mice have also demonstrated a significant dose-response relationship in multiple types of tumours, total lymphoma, lymphoma invading the organs in the mice exposed to sodium arsenate. On the other hand there were *in-vitro* studies which showed that exposure to low concentration of arsenic could be adaptive and protective

[16], causing enhanced cell proliferation and viability [17] rather than cytotoxicity. At higher concentrations toxicity of arsenic can cause an apoptotic response, as was seen in the treatment of promyelocytic leukaemia and multiple myeloma [18, 19]. This, in fact, could be a reason for not encountering tumour production in most earlier animal studies, which employed high doses of arsenicals resulting in cytotoxic effects rather than carcinogenic effects as demonstrated here.

Our results showed that the liver in mice was commonly the site of metastatic spread of haemotopoeitic tumours but not hepatocellular tumours in the exposed mice. Pregnant mice exposed to sodium arsenite during 8-18 days of gestation at doses of 42.5 and 85 mg/L in drinking water resulted in dose-dependent increases in tumours of the liver and adrenal glands in male offspring, and of the ovary in female offspring [10]. Male BALB/c mice given 3.2 mg/L arsenic in the drinking water developed fatty liver after 12 months and hepatic fibrosis after 15 months of exposure [20]. Arsenic increased hepatocellular carcinoma (14%), adenoma (23%) and total tumours (31%) compared to the control (0, 2 and 2%) in the male offspring of CD1 mice exposed to 85 mg/L arsenite during the 8th to the 18th day of gestation [21]. Hepatocellular adenoma was observed in all the test groups from the As^{V}-exposed groups. The percentage incidence in the As^{V}-exposed group was 2, 9 and 5% compared to the control 0%. In addition, liver degenerative lesions, liver cell vacuolation and fatty infiltration were seen together with preneoplastic proliferative lesions (data not shown) which could lead to cancerous endpoints. These observed lesions were similar to those seen in the liver biopsy samples of hepatomegaly from patients exposed to arsenic in Guizhou, China [22]. Our result also supports other studies where chronic arsenic exposure in the mouse produced liver cell vacuolation and fatty infiltration along with preneoplastic proliferative lesions and chronic inflammation [23, 24].

DMA was also found to be carcinogenic in the rat urinary bladder [25]. In contrast, our gross pathology showed only one bladder tumour (histiocytic sarcoma) in the 500μg As/L group, a relative low incidence of less than 2%. This suggests that sodium arsenate is not a potent carcinogen in C57BL/6J mice.

Our present study is the first long-term, low-dose carcinogenic study of arsenate which is often the major component of arsenic-contaminated drinking water found in endemic areas. It also demonstrates the dose-response relationship in carcinogenic effects of As^{V}. In the present study lung lymphoma was observed in all the treated groups and none in the control group. Histiocytic sarcoma was observed in 500 μg As/L exposed mice.

In a study by Salim et al. [26], a significant increase in tumour multiplicity (malignant lymphoma) was recorded in both $p53^{+/-}$ knockout and C57BL/6J wild type male mice exposed to 50 or 200 mg/L DMA^{V} for 18 months. The present results support the findings of these studies, that arsenate also increases the incidence of lymphomas, and with higher potency than dosing with DMA.

In conclusion, our study clearly shows that sodium arsenate is a complete carcinogen to C57Bl/6J mice, causing multi-organ tumours without any promoter. In all previous animal studies apart from that by Ng et al. [8] the doses used were relatively high compared to those associated with human exposures. This study and others [8, 10, 11] dismiss the notion that no animal model exists for arsenic carcinogenesis. These studies have also settled the debate on whether arsenic, a human carcinogen, is carcinogenic to laboratory animals. Our results can contribute to the risk evaluation of chronic arsenic exposure and the development of arsenic exposure standards.

Acknowledgement

Entox is a partnership between Queensland Health and the University of Queensland.

References

[1] IARC, Arsenic and arsenic compounds (Group 1). In: IARC monographs on the evaluation of the carcinogenic risks to humans. Supplement 7, International Agencies for Research on Cancers, Lyon, France. 1987.

[2] IARC, Monograph 84: Some drinking water disinfectants and contaminants including arsenic. International Agencies for Research on Cancers, Lyon, France. 2004.

[3] Ng, J.C., Wang, J.P. & Shraim, A., A global health problem caused by arsenic from natural sources. *Chemosphere,* 52, pp. 1353-1359, 2003.

[4] Kitchin, K.T. Recent advances in arsenic carcinogenesis: Modes of action, animal model systems, and methylated arsenic metabolites. *Toxicol. Appl. Pharmacol.,* **172**, pp. 249-261, 2001.

[5] Katnelson, B.A., Neizvestnova, Y.M. & Blokhin, V.A., Stomach carcinogenesis induction by chronic treatment with arsenic (Russian). *Vopr. Onkol.,* **32**, 68-73, 1986.

[6] Pershagen, G., Nordberg, G. & Bjorklund, N.E., Carcinomas of the respiratory tract in hamsters given arsenic trioxide and/or benzo(a)pyrene by the pulmonary route. *Environ. Res.,* **34**, pp. 227-241, 1984.

[7] Wang, J.P., Qi, L., Moore, M.R. & Ng, J.C., A review of animal models for the study of arsenic carcinogenesis. *Toxicol. Lett.,* **133**, pp.17-31, 2002.

[8] Ng, J.C., Seawright, A.A., Qi, L., Garnett, C.M., Chiswell, B. & Moore, M.R., Tumours in Mice induced by exposure to sodium arsenate in drinking water. In *Arsenic: Exposure and Health effects*, eds. C. Abernathy, R. Calderon, & W. Chappell, Elsevier Science, Oxford, London, pp. 217-223, 1999.

[9] Cui, X., Wakai, T., Shirai, Y., Shirai, Y., Hatekeyama, K. & Hirano, S., Chronic oral exposure to inorganic arsenate interferes with methylation status of p16INK4a and RASSF1A and induces lung cancer in A/J mice. *Toxicol. Sci.*, **91**, pp. 372-381, 2006.

[10] Waalkes, M.P., Ward, J.M., Liu, J. & Diwan, B.A., Transplacental carcinogenicity of inorganic arsenic in the drinking water: induction of hepatic, ovarian, pulmonary, and adrenal tumors in mice. *Toxicol. Appl. Pharmacol.,* **186 (1)**, 7-17, 2003.

[11] Waalkes, M.P., Liu, J. & Diwan, B.A., 2007. Transplacental arsenic carcinogenesis in mice. *Toxicol. Appl. Pharmacol.,* **222 (3)**, pp. 271-280, 2007.

[12] Chen, S.L., Dzeng, S.R. & Yang, M. H., Arsenic species in groundwaters of the blackfoot disease area, Taiwan. *Environ. Sci. Technol.* **28**, pp. 877-881, 1994.

[13] Krishnamohan, M., Wu, H.J., Huang, S.H., Maddelena, R., Lam, P.K.S., & Moore, M.R., Ng, J.C., Urinary arsenic methylation and porphyrin profile of C57Bl/6J mice chronically exposed to sodium arsenate. *Sci. Total Environ.* 379(2-3), pp. 235-243, 2007.

[14] Wei, M., Wanibuchi, H., Yamamoto, S., Li, W. & Fukushima, S., Urinary bladder carcinogenicity of dimethylarsinic acid in male F344 rats. *Carcinogenesis,* **20** (9), pp. 1873-1876, 1999.

[15] IPCS, Environmental Health Criteria 224: Arsenic and Arsenic Compounds. International Programme on Chemical Safety, World Health Organisation, Geneva, 2001.

[16] Snow, E.T., Sykora, P., Durham, T. R. & Klein, C.B., Arsenic, mode of action at biologically plausible low doses: What are the implications for low dose cancer risk? *Toxicol. Appl. Pharmacol.,* 207, pp. 557-564, 2005.

[17] Simeonova, P.P., Wang, S., Toriuma, W., Kommineni, V., Matheson, J., Unimye, N., Kayama, F., Harki, D., Ding, M., Vallyathan, V. & Luster, M. I., Arsenic mediates cell proliferation and gene expression in the bladder epithelium: Association with activating protein-1 transactivation. *Cancer. Res.,* **60**, pp. 3445-3453, 2000.

[18] Zheng, P.Z., Wang, K.K., Zhang, Q.Y., Huang, Q.H., Du, Y.Z., Zhang, Q.H., Xiao, D.K., Shen, S.H., Imbeaud, S., Eveno, E., Zhao, C.J., Chen, Y.L., Fan, H.Y., Waxman, S., Auffray, C., Jin, G., Chen, S.J., Chen, Z. & Zhang, J., Systems analysis of transcriptome and proteome in retinoic acid/arsenic trioxide-induced cell differentiation/apoptosis of promyelocytic leukemia. *PNAS,* **102**, pp. 7653-7658, 2005.

[19] Berenson, J. & Yeh, H., Arsenic compounds in the treatment of multiple myeloma: a new role for a historical remedy. *Clin. Lymphoma Myeloma,* 7, 192-198, 2006.

[20] Sandra, A., DasGupta, J., De, B., Roy, B. & Mazumder, D. G., Hepatic manifestations in chronic arsenic toxicity. *Ind. J. Gastroenterol.* **18**, pp. 152-155, 1999.

[21] Waalkes, M.P., Liu, J., Ward, J.M. & Diwan, B.A., Enhanced urinary bladder and liver carcinogenesis in male CD1 mice exposed to transplacental inorganic arsenic and postnatal diethylstilbestrol or tamoxifen. *Toxicol. Appl. Phamacol.,* **215**, pp. 295-305, 2006.

[22] Lu, T., Liu, J., LeCluyse, E.L., Zhou, Y.S., Cheng, M.L. & Waalkes, M.P., Application of cDNA microarray to the study of arsenic-induced liver diseases in the population of Guizhou, China. *Toxicol. Sci.* **59**, pp. 185-192, 2001.

[23] Liu, J., Liu, Y.P., Goyer, R.A., Achanzar, W. & Waalkes, M.P., Metallothionein-I-II null mice are more sensitive than wild-type mice to the

hepatotoxic and nephrotoxic effects of chronic oral or injected inorganic arsenicals. *Toxicol. Sci.,* **55**, 460-467, 2000.

[24] Waalkes, M.P., Keefer, L.K., and Diwan, B.A., Induction of proliferative lesions of the uterus, testes, and liver in Swiss mice given repeated injections of sodium arsenate: Possible estrogenic mode of action. *Toxicol. Appl. Pharmacol.,* **166**, pp. 24-35, 2000.

[25] Wei, M., Wanibuchi, H., Morimura, K., Iwai, S. & Yoshida, K., Carcinogenicity of dimethylarsinic acid in male F344 rats and genetic alterations in induced urinary bladder tumors. *Carcinogenesis,* 23, pp. 1387-1397, 2002.

[26] Salim, E.I., Wanibuchi, H., Morimura, K., Min, W., Makota, M., Yoshida, K., Endo, G. & Fukushima, S., Carcinogenicity of dimethylarsinic acid in p53 heterozygous and wild-type C57BL/6J mice. *Carcinogenesis,* **24**, pp. 335-342, 2003.

Aspects to consider for selection of chemical risk assessment methodology: the case of formaldehyde occupational exposure

S. Viegas[1,2] & J. Prista[2]
[1]*Escola Superior de Tecnologia da Saúde de Lisboa – ESTeSL/IPL (Higher School of Health Technologies of Lisbon - Polytechnic Institute of Lisbon), Portugal*
[2]*CIESP, Centro de Investigação e Estudos em Saúde Pública, Escola Nacional de Saúde Pública – ENSP/UNL (National School of Public Health – New University of Lisbon), Portugal*

Abstract

There are several risk assessment methodologies available that can be applied in contexts where occupational exposure to chemical agents occur. However, there are some aspects that should be considered for selecting a more suitable and accurate risk assessment methodology.

A study was carried out where two different risk assessment methodologies in ten anatomy and pathology laboratories were applied. One of the methodologies is propose by the Environmental Protection Agency (EPA) and the other methodology was based on the risk assessment methodology of Queensland University and defined by the authors to study this specific occupational setting.

The two risk assessment methodologies obtained different results. Application of EPA methodology for risk assessment provides data that classifies this occupational setting similar to others where occupational exposure to formaldehyde occurs. However, differences and particular characteristics of this occupational setting are not possible to know due to the fact of relying only on TWA_{8h} values. The proposal methodology ranks with high risk 30% of the activities studied in the ten laboratories and, 70% of the laboratories had at least one activity classified as high risk. The activities that were classified with very high risk and high risk were macroscopic exams developed always by the pathologist.

WIT Transactions on Ecology and the Environment, Vol 132, © 2010 WIT Press
www.witpress.com, ISSN 1743-3541 (on-line)
doi:10.2495/ETOX100031

Despite EPA methodology allowing applications in occupational settings, it only provides information about the risk for work location, not allowing a risk assessment by activity.

Keywords: risk assessment methodology, exposure assessment, occupational exposure, formaldehyde, anatomy and pathology laboratories.

1 Introduction

Nowadays people are exposed in several ways (e.g. food, drinking water, ambient air, indoor air, occupational setting) to chemical substances which are responsible for enhanced cancer risk. Often, substances previously thought to be inert or harmless to humans have been found to be carcinogenic (e.g., asbestos and vinyl chloride monomer) or toxic to the reproductive process (e.g., methylmercury and thalidomide). Moreover, an increasing number of substances have been shown to be mutagenic or carcinogenic in animal studies [1].

There are several risk assessment methodologies available and possible to apply in contexts where occupational exposure to chemical agents occur. However, there are some aspects that should be considered for a select risk assessment methodology that is more suitable and accurate because it might be followed by decisions with wide-ranging and significant consequences for workers' health and for industrial processes. All these modifications involve, usually, immense investments.

Formaldehyde, with the chemical formula CH_2O, is the most simple yet most reactive of all aldehydes. It exists as a colorless gas at room temperature and has a strong pungent smell [2, 3].

Formaldehyde is an economically important chemical with an annual production of approximately 46 billion pounds worldwide. According to the Report on Carcinogens (11th Edition, National Toxicology Program), formaldehyde ranks 25th in overall U.S. chemical production with more than 11 billion pounds produced each year [4].

Commercially, formaldehyde is manufactured as an aqueous solution called formalin, usually containing 37% by weight of dissolved formaldehyde. It is commonly used as a tissue preservative or as a bactericide in embalming procedures and in anatomy and pathology laboratories.

Given its economic importance and widespread use, many people are exposed to formaldehyde environmentally and/or occupationally. Occupational exposure involves not only individuals employed in the direct manufacture of formaldehyde and products containing it, but also those in industries utilizing these products, such as construction.

The exposed workers, commonly found in resin production, textiles or other industrial settings, inhale formaldehyde as a gas or absorb the liquid through their skin. Other exposed workers include health-care professionals, medical-lab specialists, morticians and embalmers, all of whom routinely handle bodies or biological specimens preserved with formaldehyde [5–7].

Concerning exposure limits in occupational settings, OSHA has established the following standards that have remained the same since 1992: the permissible exposure limit (PEL) is 0,75 ppm (parts per million) in air as an 8-h time-weighted average (TWA_{8h}) and the short-term (15 min) exposure limit (STEL) is 2 ppm. American Conference of Governmental Industrial Hygienists (ACGIH) recommended threshold limit value (TLV) is 0,3 ppm as a ceiling value. The National Institute of Occupational Safety and Health (NIOSH) recommends much lower exposure limits of 0,016 ppm (TWA_{8h}) and 0,1 ppm (STEL), above which individuals are advised to use respirators if working under such conditions. In Portugal, the Portuguese Norm (NP 1796 - 2007) points also 0,3 ppm as a ceiling value.

Human studies have shown that chronic exposure to formaldehyde by inhalation is associated with respiratory symptoms, and eye, nose and throat irritation [7].

Regarding the carcinogenic effects, formaldehyde was long considered as a probable human carcinogen (Group 2A chemical) based on experimental animal studies and limited evidence of human carcinogenicity [8]. However, the International Agency for Research on Cancer (IARC) reclassified formaldehyde as a human carcinogen (Group 1) in June 2004 based on ''sufficient epidemiological evidence that formaldehyde causes nasopharyngeal cancer in humans'' [5].

IARC also concluded that there was ''strong but not sufficient evidence for a causal association between leukemia and occupational exposure to formaldehyde'' [5, 7].

In relation to risk assessment, there are some articles that describe the application in occupational settings of a methodology define by Environmental Protection Agency [9]. In this case cancer risk due to the formaldehyde exposure has been assessed by estimating the excess individual lifetime cancer probability (LCP). LCP is the increase in the probability of cancer occurring.

The estimated excess LCP for formaldehyde exposure can be calculated by the equation: $R_f = C_f \times IUR_f \times L_w$, where Rf is the excess LCP for formaldehyde; Cf is the formaldehyde exposure concentration, $\mu g/m^3$; $IURf$ is the IUR factor for formaldehyde, $(\mu g/m^3)^{-1}$,which would be taken as $1{,}3\times10^{-5}$ $(\mu g/m^3)^{-1}$ [10]; and Lw is the adjustment factor for the ratio of the working time (40 years) to entire lifetime (70 years) with the value of 0,113 [11]. This methodology permits to assess cancer risk in different work locations in a specific occupational setting.

The goal of this article is to demonstrate that selection of risk assessment methodology, in the case of occupational exposure to chemical, it's necessary to consider some aspects from the chemical and also the assessment objectives.

2 Materials and methods

A study was carried out applying two different risk assessment methodologies in 10 anatomy and pathology laboratories in Portugal. One of the methodologies was already describe [9] and uses data from formaldehyde exposure assessment obtained from environmental monitoring, namely TWA_{8h} results (averages concentrations obtain in the sampling period) through an NIOSH method application (NIOSH 2541). Three exposure groups were defined, specifically pathologists, technicians and assistants and one sample for each exposure group was obtained in each laboratory (3 samples per laboratory).

The other methodology was based on a Queensland University proposal [12] that permits to perform risk assessment for each activity developed in a work station. This methodology was applied in 83 different activities developed in the 10 laboratories studied. It also used the results from environmental monitoring but, in this case, ceiling concentrations were used measured by Photo Ionisation Detection (PID) equipment (with 11.7 eV lamp), obtained in the same day of TWA_{8h} measures. Additionally, data was used from research articles about biologic adverse events associated with different formaldehyde exposure values. This data was used to categorize the health effects severity (Table 1).

Prior to methodology application, ergonomic work conditions analysis was performed to identify the different tasks developed in each work location and their execution frequency and then give the exposure probability. With these data it was possible to achieve the exposure probability (Table 2).

In order to assess the risk we multiply the likelihood of exposure by the severity categorization. The higher score gives the higher risk and define priority for applying control measures (Table 3).

Table 1: Health effects severity categorization.

Severity Categorization	Maximum concentration / effect on health associated
1. Negligible	$\leq$ 1 ppm (does not cause damage to the epithelial tissue)
2. Medium	1 $\leq$ 2 ppm (Non-neoplastic lesions of different severities and incidences)
3. Considerable	> 2 < 4 ppm (cell proliferation, metaplasia, cytotoxicity)
4. Serious	$\geq$ 4 ppm < 5 ppm (2x increase in the likelihood of nasopharyngeal cancer)
5.Very Serious	$\geq$ 5,5 ppm (4x increase in the likelihood of nasopharyngeal cancer)

Table 2: Categorization of exposure probability.

Categorization Probability	**Likelihood of Exposure**
1	Never place
2	Annually
3	Monthly
4	Weekly
5	Daily

Table 3: Scoring risk.

Score	Risk Assessment/Action
> 16	Very high risk - emerging acting
> 12 ≤ 16	High risk - Immediate response
> 6 ≤ 12	Medium risk - acting as soon as possible
> 2 ≤ 6	Low risk - No need for action, but surveillance

3 Results

Different results were obtained with the two different methodologies. With the EPA methodology all the results were above 9,2 x 10^{-3} (LCD) (Table 4).

Table 4: Results of EPA methodology.

Laboratories	Exposure Groups	Formaldehyde exposure (TWA) (ppm)	EPA methodology LCD $R_f = C_f$ x IUR_f x L_w
A	Assistants	0,27	4,8 x 10^{-4}
	Pathologists	ND	
	Technicians	0,16	2,9 x 10^{-4}
B	Assistants	0,15	2,7 x 10^{-4}
	Pathologists	0,24	4,3 x 10^{-4}
	Technician	0,16	2,9 x 10^{-4}
C	Assistants	0,12	2,2 x 10^{-4}
	Pathologists	0,47	8,5 x 10^{-4}
	Technician	0,51	9,2 x 10^{-3}
D	Assistants	ND	
	Pathologists	0,07	1,3 x 10^{-4}
	Technician	0,11	1,9 x 10^{-4}
E	Assistants	ND	
	Pathologists	0,06	1,1 x 10^{-4}
	Technician	0,07	1,2 x 10^{-4}
F	Assistants	0,09	1,6 x 10^{-4}
	Pathologists	0,23	4,1 x 10^{-4}
	Technician	0,12	2,2 x 10^{-4}
G	Assistants	0,16	2,9 x 10^{-4}
	Pathologists	0,05	8,9 x 10^{-5}
	Technician	0,04	7,2 x 10^{-5}
H	Assistants	0,25	4,5 x 10^{-4}
	Pathologists	0,11	1,9 x 10^{-4}
	Technician	0,25	4,5 x 10^{-4}

Table 4: Continued.

Laboratories	**Exposure Groups**	**Formaldehyde exposure (TWA)** (ppm)	**EPA methodology LCD** $R_f = C_f \times IUR_f \times L_w$
I	Assistants	0,05	$8,9 \times 10^{-5}$
	Pathologists	ND	
	Technician	0,06	$1,1 \times 10^{-4}$
J *	Assistants	0,13	$2,3 \times 10^{-4}$
	Pathologists	0,08	$1,4 \times 10^{-4}$

* Do not have assistants working in the laboratory.
ND – Not detectable.

Concerning the proposal methodology results, they have different distribution between the laboratories studied (Figure 1).

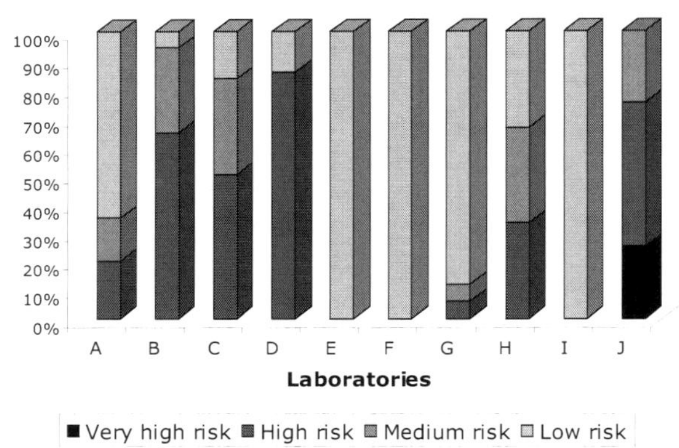

Figure 1: Results of proposal methodology.

Laboratories E, F and I have all the activities classified with low risk. Laboratory D have 86% of the activities classified with high risk.

Concerning the risk classification distribution per activity, 2,41% have very high risk classification, 32,53% obtained the high risk classification, 13,25% were classified with medium risk and, finally, 51,81% have low risk classification.

We could also conclude that 30% of the laboratories have all activities classified with low risk and 70% of the laboratories have at least one activity classified with high risk. The activities that were classified with very high risk and high risk were macroscopic exams developed always by the pathologist.

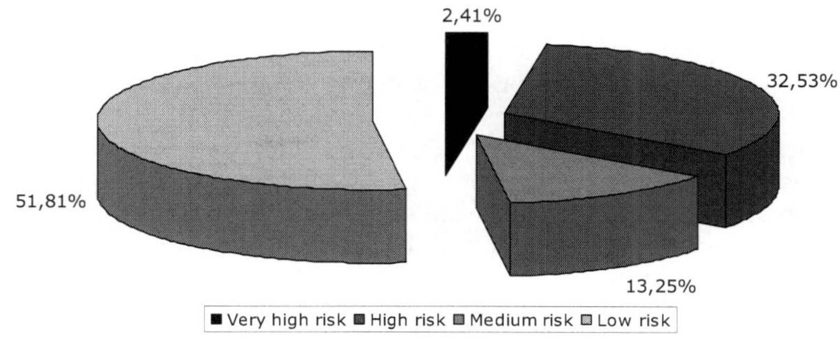

Figure 2: Results of risk classification distribution per activity.

4 Discussion

For some genotoxic carcinogens the existence of a "practical" threshold is supported by studies on mechanisms and/or toxicokinetics. Formaldehyde is one of the chemicals and, therefore, a NOAEL (No Observed Adverse Effects Level) may be established from which to derive a health-based exposure limit [13-16]. Considering these characteristics it was possible to propose this new risk assessment methodology, based on Queensland University proposal and make an association between occupational exposure to formaldehyde air concentrations and health effects.

Recent studies [17, 18] showed EPA methodology application in occupational settings with FA exposure but, as suggested by the methodology described, making use of the TWA_{8h} values obtained in the situations studied. Thus, applying this equation to ours results we obtain values that are lower and equal to $9,3 \times 10^{-3}$ (LCP) when, the cancer risk from formaldehyde exposure in general population is 1×10^{-6} LCP, and in occupational settings, will be greater than 1×10^{-4} LCP [11,19].

We conclude that application of EPA methodology for risk assessment provides data that classifies this occupational setting similar to others where occupational exposure to formaldehyde occurs. However, differences and particular characteristics of this occupational setting are not possible to know due to the fact of relying only on TWA_{8h} values, appointed as less appropriate with regard to assess formaldehyde occupational exposure [5,20].

Despite EPA methodology also allow application in occupational settings, provides only information about the risk for work location, as performed in the study of He and Zhang (2009), not allowing a risk assessment by activity.

Occupational health interventions highlight the importance of knowing the most critical activities because permits intervention prioritization and identification of technical and/or organizational measures aiming to minimize and/or eliminate exposure (to know which activity has a greater contribution to exposure and the constraints of activity, allow knowing the variables that influence the exposure).

Moreover, this kind of information provides important information for raising awareness for exposure prevention (which activity requires protective measures to be strengthened and/or employed by workers) and, last but not least, a risk assessment more detailed, allowing also identifying professional group with the most critical exposure "anticipating" potential effects on health through the adequacy of health vigilance activities. The proposal methodology gives this information allowing more complete and meticulous interventions. In this case the macroscopic exam was the task with higher risk and the pathologist group with the higher exposure.

Finally, many environmental and occupational chemicals, toxicants and carcinogens require metabolic activation to exert their action. However, metabolic polymorphisms can modulate individual response [21]. Also, a consistent, positive association of DNA repair deficiency and increased risk was recently shown by an extended review of inter-individual variability in DNA repair systems and cancer risk [22].

Taking into account these aspects we have to mention that despite the important and useful information that both methodologies gives, there is no consideration about individual variability concerning with the capacity of dealing with a specific chemical exposure.

5 Conclusions

For occupational health interventions it's important to know the activities that increment exposure and the workers group with the higher exposure to define more adequate and successful preventive and protective measures. So, when selecting a risk assessment methodology aiming at occupational health interventions we have to consider these aspects.

In the case of formaldehyde occupational exposure in anatomy and pathology laboratories it seems that "macroscopic exams" is the worst activity concerning exposure and the "pathologists" the workers group with the higher exposure.

In conclusion, all risk assessment methodologies have limitations that have to be considered and known permitting a better methodology selection. Also, obtained data showed that occupational exposure to formaldehyde in anatomy and pathology laboratories in Portugal is still a matter of great concern.

References

[1] Herber, R, *et al.*, Risk assessment for occupational exposure to chemicals. A review of current methodology. *Pure Applied Chemistry.* 73 (2001) 993 – 1031.

[2] IPCS – INTERNATIONAL PROGRAMME ON CHEMICAL SAFETY - *Human exposure assessment: concepts and principles*. Geneva: WHO, 1993 (Environmental Health Criteria; 155).

[3] Goyer, N, *Exposition au Formaldéhyde en Milieu de Travail : La Pathologie*. Montréal: Institut de Recherche Robert-Sauvé en Santé et en Sécurité du Travail. Fiche technique RG3-471, Montréal, IRSST, 2007.

[4] NTP, National Toxicology Program: Formaldehyde (Gas) CAS no. 50-00-0, 2005. http://ntp.niehs.nih.gov/ntp/roc/eleventh/profiles/s089form.pdf.
[5] IARC, *Formaldehyde, 2-Butoxyethanol and 1-tert-Butoxypropan-2-ol*. Lyon: International Agency For Research on Cancer, 2006. (IARC Monographs on the Evaluation of Carcinogenic Risks to Humans; 88).
[6] Vincent, R, Jandel, B, Exposition professionnelle au formaldéhyde en France : informations fournies par la base de données Colchic. *Hygiène et sécurité du travail. Cahiers de notes documentaires.* (2e trimestre 2006) 19-33.
[7] Zhang, L, *et al.*, Formaldehyde exposure and leukemia: A new meta-analysis and potential mechanisms. *Mutation Research/Reviews in Mutation Research.* 681 (2009) 150-168.
[8] Binetti, R, Costamagna, F, Marcello, I, Development of carcinogenicity classifications and evaluations: the case of formaldehyde. *Annali Istituto Superiore di Sanitá.* 42 : 2 (2006) 132-143.
[9] US EPA, *Guidelines for Carcinogen Risk Assessment*, Federal Regulation, United States Environmental Protection Agency 51, (1986) 33992–35003.
[10] US EPA Integrated Risk Information System, http://www.epa.gov. Washington, DC, United States Environmental Protection Agency, 1998.
[11] Wu, P, Li, Y, Lee, C, Chiang, C, Su, H, Risk assessment of formaldehyde in typical office buildings in Taiwan. *Indoor Air.* 13 (2003) 359-363.
[12] University of Queensland, Occupational health and safety risk assessment and management guideline, 2005. http://www.uq.edu.au/hupp/?page=25024&pid=25015 (27.10.2009)
[13] Morgan, K, Review article: A brief review of formaldehyde carcinogenesis in relation to rat nasal pathology and human health risk assessment. *Toxicologic Pathology*. 25 (1997) 291 – 307.
[14] Bolt, H.; Degen, G. – Human carcinogenic risk evaluation, Part II: Contributions of the Eurotox specialty section for carcinogenesis. *Toxicological Sciences.* 81 (2004) 3-6.
[15] Bolt, H, Huici-Montagud, A, Strategy of the scientific committee on occupational exposure limits (SCOEL) in the derivation of occupational exposure limits for carcinogens and mutagens. *Archives of Toxicology.* 2007.
[16] Henglester, J, *et al.*, Challenging Dogma: Threshold for genotoxic carcinogens? The case of vinyl acetate. *Annu. Rev. Pharmacol. Toxicol.* 43 (2003) 485-520
[17] He, Z, Zhang, Y, Risk assessment of exposure to formaldehyde in a medium density fibreboard plant in China. In *Healthy Buildings 2009*, 13 to 17 of September, United States, 2009.
[18] Pillidis, G, *et al.*, Measurements of benzene and formaldehyde in a medium sized urban environment. Indoor/outdoor health risk implications on special populations groups. *Environ. Monit. Assess.* 150 (2009) 285–294.
[19] Gratt, L, *Air toxic risk assessment and management.* New York. Van Nostrand Reinhold, 1996.

[20] Pyatt, D, Natelson, E, Golden, R, Is inhalation exposure to formaldehyde a biological plausible cause of lymphohematopoietic malignancies? *Regulatory Toxicology and Pharmacology*. 51 (2008) 119-133.
[21] Kelada, S, et al., The role of genetic polymorphisms in environmental health. *Environmental Health Perspectives*. 111 (2003) 1055-1064.
[22] Berwick, M, Vinéis, P, Markers of DNA repair and susceptibility to cancer in humans: an epidemiologic review. *J. Natl. Cancer Inst*. 92 (2000) 847-897.

The possibility of removal of endocrine disrupters from paper mill waste waters using anaerobic and aerobic biological treatment, membrane bioreactor, ultra-filtration, reverse osmosis and advanced oxidation processes

D. Balabanič[1, 3], D. Hermosilla[2], A. Blanco[2], N. Merayo[2]
& A. Krivograd Klemenčič[4]
[1]Pulp and Paper Institute, Slovenia
[2]Chemical Engineering Department, Complutense University of Madrid, Spain
[3]University of Nova Gorica, Slovenia
[4]University of Ljubljana, Slovenia

Abstract

An endocrine disrupter is an exogenous agent that interferes with the synthesis, binding, secretion, transport, action or elimination of natural hormones in the body that are responsible for the maintenance of homeostasis, reproduction, development, and behaviour. Some of them are suspected of causing abnormalities in sperm and increasing hormone-related cancers in humans. Studies have also been published on the estrogen-like responses of endocrine disrupters in wildlife, such as birds, amphibians, reptiles and fish. Endocrine disrupters include a wide variety of pollutants such as alkylphenols, bisphenol A, pesticides, polycyclic aromatic hydrocarbons (PAHs), phthalates, heavy metals, and natural or synthetic hormones. They may be released into the environment in different ways. One of the most important sources of endocrine disrupters are industrial waste waters. The conventional waste water treatment processes are not specifically designed to remove traces of dangerous organic contaminants (except for heavy metals) so the latter are consequently consumed by aquatic organisms and through them may also enter human food chain. In the presented research the following treatments for removing of organic endocrine disrupting

WIT Transactions on Ecology and the Environment, Vol 132, © 2010 WIT Press
www.witpress.com, ISSN 1743-3541 (on-line)
doi:10.2495/ETOX100041

compounds from paper mill waste waters were compared: anaerobic biological treatment, membrane bioreactor, and reverse osmosis (pilot plant A), and combined (anaerobic and aerobic) biological treatment, ultra-filtration and reverse osmosis (pilot plant B) at pilot scale and advanced oxidation processes (Fenton, photo-Fenton, photo-catalysis with TiO_2 and ozonation) at laboratory scale. The results indicated that the concentrations of organic endocrine compounds from paper mill waste waters were efficiently reduced (100%) by both combinations of pilot plants, photo-Fenton oxidation (95%) while the ozonation, photo-catalysis with TiO_2 reagent and Fenton reaction was less effective (70–80%).

Keywords: endocrine disrupting compounds, anaerobic treatment, aerobic treatment, membrane filtration, advanced oxidation processes, waste water treatment.

1 Introduction

Endocrine disrupting chemicals (EDCs) became the focus of both public and scientific interest when defects in sexual behaviour and reproductive ability of wild-living animals were ascribed to their steroid-like and anti-steroid androgenic properties. These chemicals include different groups of compounds, such as alkylphenol compounds, bisphenol A, dioxins, polyaromatic hydrocarbons (PAHs), polychlorinated bisphenyls (PCBs), phthalates, pesticides, and heavy metals such as cadmium, lead or mercury [1–4]. Environmentally detrimental chemicals with endocrine activity have effect in human health and most of them are mutagenic and highly carcinogenic [5–7].

EDCs may be released into the environment in different ways and one of the most important sources is industrial wastewater. The papermaking industry is not an exception. The pulp and paper industry is the sixth largest polluter, discharging a variety of liquid, solid and gaseous wastes into the environment [8]. The main environmental issues are emissions to water and energy consumption. It is the pollution of water bodies, however, which is of major concern because large volumes of wastewater are generated for each metric ton of paper produced, depending on the raw material, finished product and extent of water reuse. Untreated paper mill effluent discharges cause considerable damage to the receiving waters, since they have high chemical oxygen demand (COD), biochemical oxygen demand (BOD), alkylphenolic and chlorinated compounds, suspended solids (mainly fibres), bisphenol A, phthalates, resin acids, lignin and its derivatives [9–12].

Paper mill effluents will have to be more strictly controlled in order to preserve natural balance in the environment. According to the EU Water Framework Directive 2000/60/EC [13], all industrial water pollution sources will have to be regularly analysed for the content of numerous compounds which are toxic, bio-accumulative or they have function as endocrine disruptors [9, 10]. Consequently, individual producers will be obliged to reduce the impact of their discharges in order to fulfil the requirements of the directive. These necessary steps will ensure better environmental management all around Europe and will also help restore already highly damaged equilibrium of natural surface and underground waters.

Conventional waste water treatment processes are not specifically designed to remove traces of EDCs (except heavy metals), so the latter are consequently consumed by aquatic organisms, and then represent a hazard to the whole food chain. For this reason, it is essential that future research focus on the investigation of appropriate treatment methods that can be integrated into water and waste water treatment facilities to prevent the release of EDCs into the natural waters. Particularly, membrane technologies (ultra-filtration, nano-fitration, and reverse osmosis) and advanced oxidation processes (AOPs) have recently shown promising results in the removal of pollutants from waste waters and surface waters [14].

Activated sludge systems have been successfully applied to treat a wide variety of waste waters, and more than 90% of the municipal and industrial wastewater treatment plants use this treatment type as an important part of their treatment train. Several microorganisms, including bacteria, fungi and yeasts are known for their ability to degrade hydrocarbons to CO_2, H_2O and bacterial cells [15]. Biological treatment, particularly by activated sludge process, has been widely used for the removal of organic compounds from paper mill wastewaters [16].

Membrane filtration technologies, such as ultra-filtration (UF), nano-filtration (NF), and reverse osmosis (RO) have been shown as a promising alternative for removing micro-pollutants [17]. RO will provide and almost complete removal, but the higher implied energy consumption is an important drawback to be considered. Some studies have shown that membrane bioreactors (MBRs) could remove more than 80% of organic potential EDCs from wastewater [18].

There is an increasing interest in utilization of AOPs for destruction of slow degrading compounds. AOPs are based on the production of hydroxyl radicals (OH·) as oxidizing agents to minimize the complex chemicals in the effluents. Fenton reagent (hydrogen peroxide and ferrous iron) is a relatively cheap and easy to operate treatment in comparison to other advanced oxidation processes [19]. Many studies have shown that the oxidising power of Fenton process can be greatly enhanced by combination with the irradiation of UV [20].

Photo-catalysis is another chemical oxidation process in which a metal oxide semiconductor immersed in water and irradiated by near UV light results in the formation of free hydroxyl radical (OH·). TiO_2 is the most widely used catalyst, mainly because of its photo stability, non-toxicity, and water insolubility under most environmental conditions [21].

Ozone can be used for treatment of effluents from various industries [22]. A major disadvantage of the ozonation is the relatively high cost of ozone generation coupled with the short half-life ozone period. Thus, ozone always needs to be generated at site. Some studies indicating that the ozonation was highly efficient for removal organic pollutants and decolorization [16].

The purpose of this study was to investigate, among anaerobic and aerobic biological treatment, membrane filtration trains and AOPs, the most effective method for removing organic endocrine disrupting compounds from a paper mill waste water. Results will provide useful information to industrial applications.

2 Materials and methods

2.1 Material and analytical methods

Effluent samples from a paper mill, before and after every step of the pilot plant A and pilot plant B, were collected in 3L glass bottles. All analyses were made according to the Standard Methods for the Examination of Water and Wastewater (APHA, AWWA and WPCF, 2005) [23]. All chemicals used were of analytical grade.

Organic EDCs were extracted from the samples with use solid phase extraction (SPE) cartridges (Oasis HLB and Strata-X). An Agilent 7890 GC-MS (gas chromatography-mass spectrometry) system with autosampler was used for the determinations of organic EDCs.

Different treatments were compared in their removal efficiency of organic EDCs in the collected wastewater, comprising anaerobic treatment, aerobic treatment, MBR, UF, and RO, at a pilot plant scale, and the following advanced oxidation processes (AOPs) were performed at laboratory scale: Fenton, photo-Fenton, ozonation and photo-catalysis with TiO_2.

2.1.1 Pilot plants treatments

Two pilot plants were installed in the paper mill for the treatment of its waste water flowing out from the dissolved air flotation system placed in the first water loop of the paper mill, which is actually the most contaminated water of the mill.

Pilot plant A consisted of an anaerobic reactor followed by a MBR and a final RO filtration. The effluent of the anaerobic reactor of pilot plant A was discharged into MBR of PES hollow-fibre membranes of 0.05 μm nominal pore size. The water from the MBR entered then into the RO section, formed from spiral wound membranes of polyamide material.

Pilot plant B consisted of a biological double step (anaerobic + aerobic) followed by UF and RO filtration. The effluent of the anaerobic reactor in pilot plant B is discharged into an activated sludge reactor, divided in three successive cascade basins, through where the water flow in series. An aeration system continuously supplied oxygen to the waste water. A secondary sedimentation serves for the separation of the activated sludge from the wastewater coming out from the biological treatments stage. The settled sludge is controlled with a return sludge pumping station which drives it back periodically to the aerobic tank. In this way, the concentration of the activated sludge is kept as constant as possible. The clarified water is then sent to an UF membrane system. UF membranes removed particles greater than 0.04 μm. Finally, the RO unit was formed from membrane modules of spiral wounded polymer with a pore size in the range of 0.1nm. These modules were operated in cross-flow conditions.

2.1.2 AOPs treatments

The same waste water from the paper mill effluent was collected as described above and treated, in the laboratory, evaluating EDCs removal, by the following AOPs: conventional Fenton, photo-Fenton, ozonation and photo-catalysis with TiO_2.

2.1.2.1 Conventional Fento n The experiments were performed in a 3L reactor where 2L of the paper mill waste water were placed and mixed throughout every experiment with a magnetic device. The temperature was adjusted to 20ºC with a water heater and circulator. The pH was continuously and automatically adjusted to 3 (±0.1) along the treatment using 1 mol·L^{-1} sulphuric acid (H_2SO_4) or 1 mol·L^{-1} sodium hydroxide (NaOH). These temperature and pH values are reported to produce best results in the removal of contaminants in waste waters [24]. After temperature and initial pH adjustment, ferrous sulphate ($FeSO_4$) was added to reach the targeted ferrous ion concentration. Five different concentrations of Fe^{2+} were tested, corresponding to $[H_2O_2]:[Fe^{2+}]$ ratios. Hydrogen peroxide (H_2O_2) was then added in batch mode until the designed concentration was reached.

Before the start of the experiment and to the time intervals 20, 40, 60, 80, 100 and 120 minutes, treated water was taken to analysed it. Before extracted with SPE all samples were filtered through a 0.45 μm filter.

2.1.2.2 Photo-Fenton treatment In all these photo-Fenton treatment experiments, the experimental protocol was the same as the one described above for conventional Fenton experiments despite that a 450 W high-pressure mercury immersion lamp was used. This lamp was enclosed inside a quartz glass vessel through which water was circulated in order to reduce the excessive heat generated during the UV irradiation. The lamp was located vertically in the centre of the reactor. The entire assembly was introduced in a safety cabinet.

2.1.2.3 Ozonation The ozonation system included oxygen gas and ozone generator, ozone bubble and trap reactors. The pure oxygen from oxygen generator was allowed to pass into the ozone generator and them the diffuser connecting to the bottom of ozone bubble reactor created bubbles in the ozone reactor filled with paper mill effluent. The ozone flow rate was 4 L/min, with ozone concentration 0.5, 1.5 and 3 ppm. The reaction temperature was 20 °C during all experiments.

2.1.2.4 Photo-catalysis wit h TiO_2 Calculated concentration of TiO_2 was added to the paper mill wastewater and placed into the safety cabinet were the UV light was irradiated with the 450 W lamp described above along 180 min. 200 ml samples were collected and filtered at 0.45 μm every 30 min for measuring EDCs.

3 Results and discussion

In our study we found that the combination of biological treatment and membrane filtration are very effective (100%) for removal EDCs from paper mill waste water (Fig. 1-2). RO provided totally removal presented organic EDCs, but the higher implied energy consumption is an important drawback to be considered. The highest removal efficiency of photo-catalysis with TiO_2 reagent, conventional Fenton reaction and ozonation is 70-85% (Fig. 3-5).

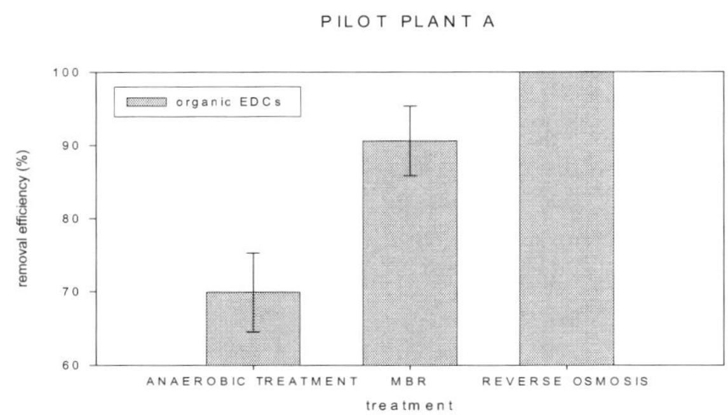

Figure 1: Removal efficiency along the pilot plant A.

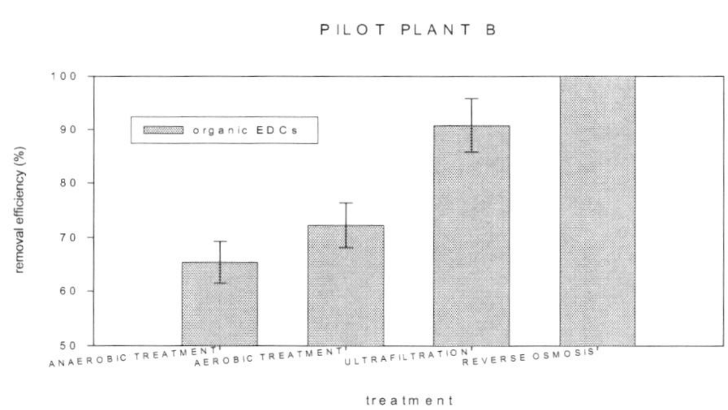

Figure 2: Removal efficiency along the pilot plant B.

Figure 4 shows the EDCs removal efficiency of the performed conventional Fenton treatments to paper mill wastewater, differing on the added concentration of Fe^{2+}. By adding of Fe^{2+}, the removal efficiency of organic substances increased until H_2O_2:Fe^{2+} ratios 2.

Ozone concentration 1.5 ppm with ozone flow rate 4 L/min is enough for 80% EDCs removal from paper mill waste waters (Fig. 5).

Fig. 6 shows that the most effectiveness AOP is photo-Fenton reaction with 95% removal efficiency.

The highest removal efficiency among AOPs was photo-Fenton reaction (Fig. 7).

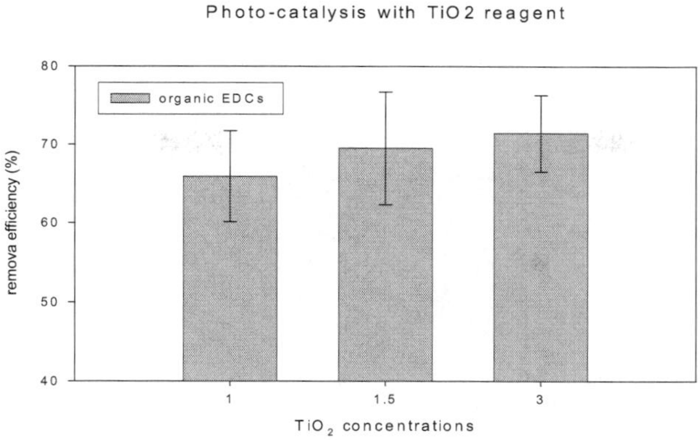

Figure 3: Removal efficiency of photo-catalysis with TiO_2 reagent according the TiO_2 concentrations.

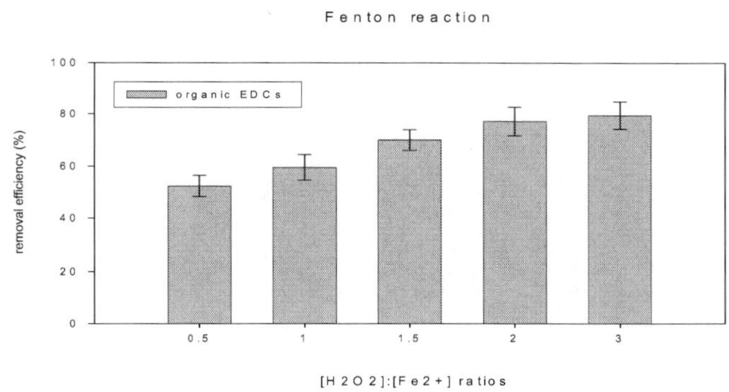

Figure 4: Removal efficiency of conventional Fenton reaction according the H_2O_2:Fe^{2+} ratios.

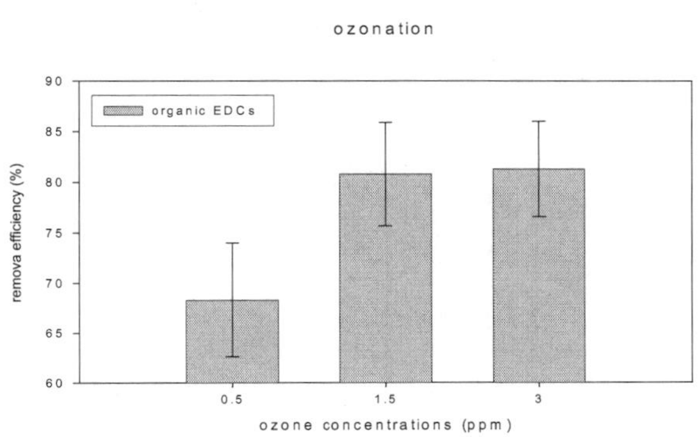

Figure 5: Removal efficiency of ozonation according the ozone concentrations.

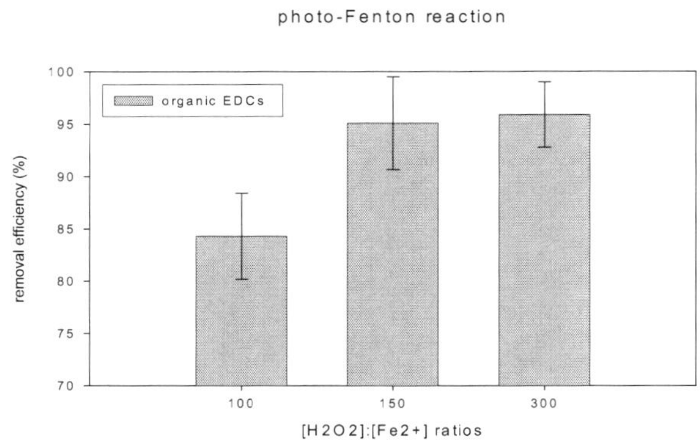

Figure 6: Removal efficiency of photo-Fenton reaction according the H_2O_2:Fe^{2+} ratios.

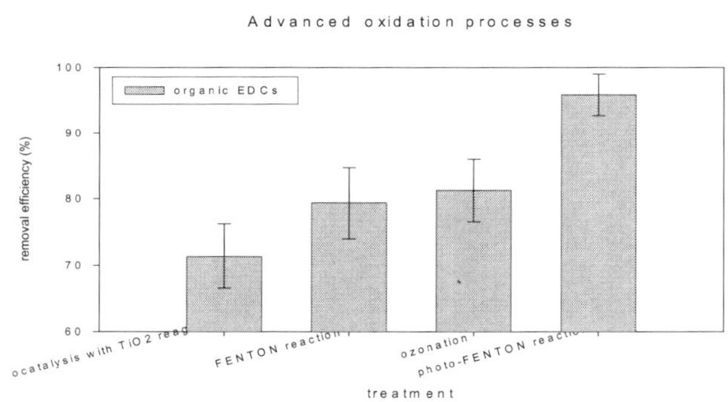

Figure 7: Removal efficiency of AOPs.

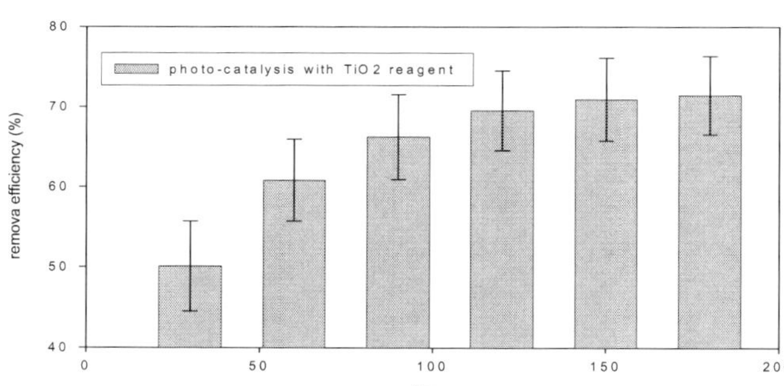

Figure 8: Removal efficiency of photo-catalysis with TiO_2 reagent according to the experiment duration.

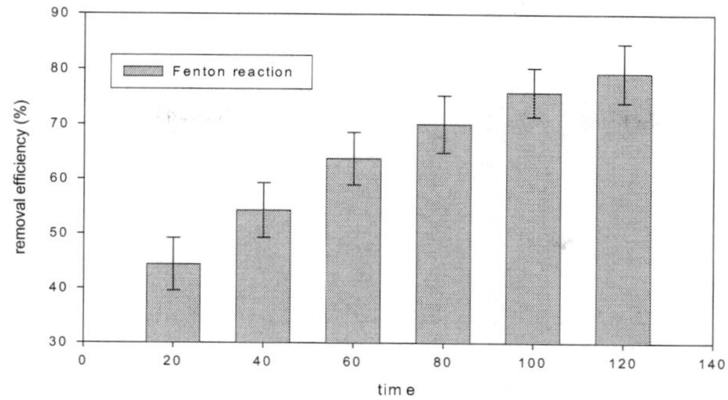

Figure 9: Removal efficiency of Fenton reaction according to the experiment duration.

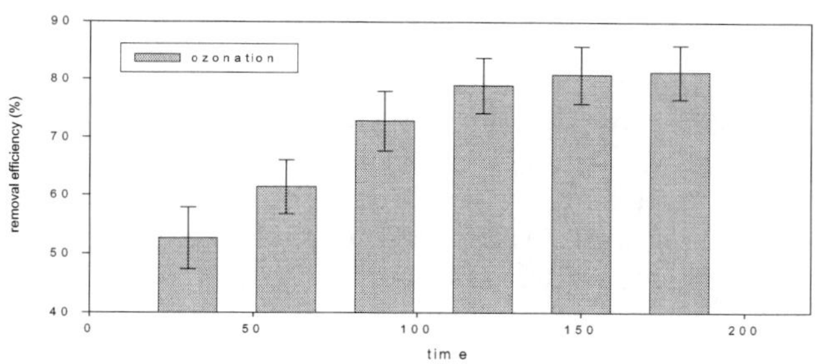

Figure 10: Removal efficiency of ozonation according to the experiment duration.

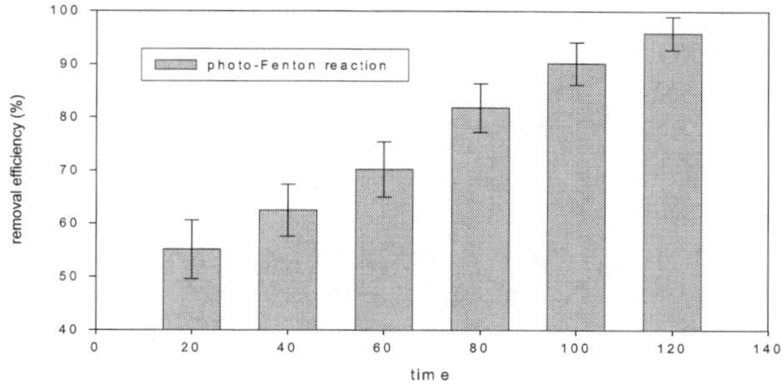

Figure 11: Removal efficiency of photo-Fenton reaction according to the experiment duration.

All experiments with AOPs treatment shows 120 min as enough time to achieve maximum removal effectiveness. (Figs. 8–11).

Figure 7 shows that Fenton reaction has been over a 15% more efficient in the presence of UV light (photo-Fenton reaction). Despite the high efficiency of photo-Fenton reaction, the main disadvantage of this method is the larger power consumption.

4 Conclusions

In the presented study, we investigated different possibilities to remove organic EDCs from the paper mill waste waters. Since some substances present in the paper industry effluents are harmful to the environment and organisms in very low concentrations, such as EDCs, it will be necessary to prevent, or at least limit, their input in the production process; or otherwise perform successful treatments to their removal. Our study shows that the combination of biological treatment and membrane filtration are 100% effective for removal EDCs from paper mill waste water. Reverse osmosis provided totally removal presented organic EDCs, but the higher implied energy consumption is an important disadvantage to be considered. Paper mills are one of the largest industrial consumers of water, and they will have to adopt new technologies of waste water treatment in the future, so they may provide a better water quality, and thus, contribute significantly to a cleaner environment and safer future.

Results showed that among the selected methods of organic contamination treatment of paper mill waste waters, RO, photo-Fenton, and MBR were the most efficient on removing organic EDCs.

Acknowledgements

This work was supported by the Slovenian Technology Agency, and it was developed in the framework of the projects: "PROLIPAPEL" (S-0505/AMB-0100), funded by the Regional Government of Madrid (Comunidad Autónoma de Madrid), Spain; "AGUA Y ENERGÍA" (CTM2008-06886-C02-01), funded by the Ministry of Science and Innovation (Ministerio de Ciencia e Innovación) of Spain; and "AQUAFIT4USE" (211534), funded by the European Union.

References

[1] Evans, N.P., North, T., Dye, S. & Sweeney, T., Differential effects of the endocrine-disrupting compounds Bisphenol-A and Octylphenol on gonadotropin secretion, in prepubertal ewe lambs. *Domestic Animal Endocrinology,* **26(1)**, pp. 61-73, 2004.

[2] Folmar, L.C., Hemmer, M.J., Denslow, N.D., Kroll, K., Chen, J., Cheek, A., Richman, H., Meredith, H. & Grau, E.G., A comparison of the estrogenic potencies of estradiol, ethynylestradiol, diethylstilbestrol,

nonylphenol and methoxychlor in vivo and in vitro. *Aquatic Toxicology,* **60(12)**, pp. 101-110, 2002.

[3] Ishihara, A., Nishiyama, N., Sugiyama, S. & Yamauchi, K., The effect of endocrine disrupting chemicals on thyroid hormone binding to Japanese quail transthyretin and thyroid hormone receptor. *General and Comparative Endocrinology,* **134(1)**, pp. 36-43, 2003.

[4] McKinlay, R., Plant, J.A. & Bell, J.N., Voulvoulis N., Endocrine disrupting pesticides: Implications for risk assessment. *Environment International,* **34(2)**, pp. 168-183, 2008.

[5] Birkett, J.W., *Endocrine Disrupters in Wastewater and Sludge Treatment Processes,* IWA Publishing, Lewis Publishers: London, pp. 1-34, 2003.

[6] Petrovic, M., Eljarrate, E., Lopez De Alda, M.J. & Barcelo, D., Endocrine disrupting compounds and other emerging contaminants in the environment: A survey on new monitoring strategies and occurence data. *Analytical and Bioanalytical Chemistry,* **378(3),** pp. 549-562, 2004.

[7] Rai, U.N. & Pal, A., Health Hazards of Heavy Metals. http://isebindia.com/01_04/02-01-2.html.

[8] Ali, M. & Sreekrishnan, T.R., Aquatic toxicity from pulp and paper mill effluents: a review. *Advances in Environmental Research,* **5(2)**, pp. 175-196, 2001.

[9] Hamm, U., Oller, H.J. & Kuwan, K., Endokrine Substanzen in Abwassern der Papierindustrie. *IPW*, **1**, pp. 45-48, 2005a.

[10] Hamm U, Oller HJ, Kuwan K. Endokrine Substanzen in Abwassern der Papierindustrie (II). *IPW*, **2**, pp. 47-49, 2005b.

[11] Kimura, K., Hara, H. & Watanabe, Y., Elimination of Selected Acidic Pharmaceuticals from Municipal Wastewater by an Activated Sludge System and Membrane Bioreactors. *Environmental Science and Technology,* **41(10)**, pp. 3708-3714, 2007.

[12] Wang, Y.Q., Hu, W., Cao, Z.H., Fu, X.Q. & Zhu, T., Occurence of endocrine-disrupting compounds in reclaimed water from Tianjin, China. *Analytical and Bioanalytical Chemistry,* **383(5)**, pp. 857-863, 2005.

[13] Directive 2000/60/EC of the European Parliament and of the Council of 23 October 2000, establishing a framework for Community action in the field of water policy. Official Journal of the European Communities.

[14] Gültekin, I. & Ince, N.H., Synthetic endocrine disruptors in the environment and water remediation by advanced oxidation processes. *Journal of Environmental Management,* **85(4)**, pp. 816-832, 2007.

[15] Shokrollahzadeh, S., Azizmohseni, F., Golmohammad, F., Shokouhi, H. & Khademhaghighat, F., Biodegradation potential and bacterial diversity of a petrochemical wastewater treatment plant in Iran. *Bioresource Technology,* **99(14)**, pp. 6127-6133, 2008.

[16] Pokhrel, D. & Viraraghavan, T., Treatment of pulp and paper mill effluent–a review. *Science of the Total Environment,* **333(1-3)**, pp. 37-58, 2004.

[17] Yoon, Y., Westerhoff, P., Snyder, S.A. & Wert, E.C., Nanofiltration and ultrafiltration of endocrine disrupting compounds, pharmaceuticals and

personal care products. *Journal of Membrane Science,* **270(1-2)**, pp. 88-100, 2006.
[18] Wintgens, T., Gallenkemper, M. & Melin, T., Endocrine disrupter removal from wastewater using membrane bioreactor and nanofiltration technology. *Desalination,* **146(1-3)**, pp. 387-391, 2002.
[19] Arslan-Alaton, I., Gursoy, B.H. & Schmidt, J.E., Advanced oxidation of acid and reactive dyes: Effect of Fenton treatment on aerobic, anoxic and anaerobic processes. *Dyes Pigments,* **78(2)**, pp. 117-130, 2008.
[20] Sun, J.H., Sun, S.P., Fan, M.H., Guo, H.Q., Lee, Y.F. & Sun, R.X., Oxidative decomposition of p-nitroaniline in water by solar photo-Fenton advanced oxidation process. *Journal of Hazardous Materials,* **153(1-2)**, pp. 187-193, 2008.
[21] Li, Y., Li, X., Li, J. & Yin, J., Photocatalytic degradation of methyl orange by TiO2-coated activated carbon and kinetic study. *Water Research,* **40(6)**, pp. 1119-1126, 2006.
[22] Baban, A., Yediler, A., Lienert, D., Kemendere, N. & Kettrup, A., Ozonation of high strength segregated effluents from a woollen textile dyeing and finishing plant. *Dyes and Pigments,* **58(2)**, pp. 93-98, 2003.
[23] APHA, AWWA, WPCF. Standard methods for the examination of water and wastewater: Washington DC, 2005.
[24] Amat, A.M., Arqués, A., López, F., Miranda, M.A., Solar photo-catalysis to remove paper mill wastewater pollutants. *Solar Energy,* **79(4)**, pp. 393-401, 2005.

Poultry fungal contamination as a public health problem

C. Viegas[1], C. Veríssimo[2], L. Rosado[2] & C. Silva Santos[3]
[1]*Higher School of Health Technologies of Lisbon, Polytechnic Institute of Lisbon, Portugal*
[2]*National Institute of Health Dr. Ricardo Jorge, Mycology Laboratory, Portugal*
[3]*School of Public Health, New University of Lisbon, Portugal*

Abstract

A descriptive study was developed to monitor poultry fungal contamination. Five air samples of 100 litres through impaction method were collected and 4 swab samples from surfaces were also collected using a 10 cm square of metal. Simultaneously, temperature and humidity were monitored as well. Twenty different species of fungi in air were identified, being the 4 most commonly isolated the following genera: *Cladosporium* (40,5%), *Alternaria* (10,8%), *Chrysosporium* and *Aspergillus* (6,8%). Concerning surfaces, 21 different species of fungi were identified, being the 4 genera more identified *Penicillium* (51,8%), *Cladosporium* (25,4%), *Alternaria* (6,1%) and *Aspergillus* (4,2%). In addition, *Aspergillus flavus* also isolated in the poultry air, is a well-known producer of potent mycotoxins (aflatoxins) and besides this species other isolated genera, like *Fusarium* and *Penicillium,* are also known as mycotoxins producers. Also noteworthy is the fact that *Aspergillus fumigatus*, one of the species isolated in air and surfaces, is one of the saprophytic fungi most widespread in air and is capable of causing severe or sometimes fatal aspergillosis. There was no significant relationship ($p>0,05$) between fungal contamination and temperature and humidity.

Keywords: poultry, fungal contamination, mycotoxins, public health problem.

1 Introduction

Fungi presence requires ideal conditions of temperature, humidity, oxygen, carbon sources, nitrogen and minerals. Their biological activities of

WIT Transactions on Ecology and the Environment, Vol 132, © 2010 WIT Press
www.witpress.com, ISSN 1743-3541 (on-line)
doi:10.2495/ETOX100051

biodegradation and biodeterioration, depend on their enzymes activity, the environmental conditions, the competition phenomenon and the nature of the substrate. In situations where the fungal concentrations are high or when people suffer from respiratory problems or have a weak immune system, exposure to fungi can cause the onset of symptoms and disease. The effects are dependent on the species present, the metabolic products, the concentration and exposure duration and individual susceptibility [1].

Until now, epidemiological studies have failed to establish a causal relationship of the extent of fungal presence, exposure time and specific effects on health or frequency and severity of symptoms reported. Studies tend to show only existence of a link between exposure to fungi and development of symptoms, especially respiratory symptoms [1]. However, fungal species are generally identified as the cause of allergic diseases, headaches, eye irritation, obstruction of airways, coughing and other symptoms [2].

In addition, a group of indoors molds can produce secondary metabolites, like mycotoxins, in response to changes in their environment. Mycotoxins can be pro-inflammatory, immunosuppressive or carcinogenic [3]. The different chemical groups of mycotoxins include aflatoxins, fumonisins, ochratoxins, rubratoxins, and trichothecene toxins, all with different biologic properties [3]. Agricultural operations, such as animal feeding, increase farmers' risk of exposures to airborne dust and micro organisms like fungi [4]. Besides that, in Portugal there is an increasingly industry of large facilities that produce whole chickens for domestic consumption. Although much research has been done on microbial contaminants associated with the various stages of processing poultry and meat products [5, 6], only few investigations have reported on the indoor air of these plants [7, 8]. Moreover, air plays a significant role in the poultry meat contamination [7] and there is evidence that mycotoxins can cause human disease from the ingestion of fungus-contaminated food [9].

This investigation was designed to describe in one poultry environmental fungal contamination phenomena and explore possible associations with independent environmental variables.

2 Materials and methods

A descriptive study was developed to monitor one poultry fungal contamination. Five air samples of 100 litres each through impaction method were collected and 4 swab samples from surfaces were also collected using a 10 cm square of metal. Simultaneously, two environmental parameters – temperature and relative humidity – were monitored, using the Babouc equipment, (LSI Sistems), according to the International Standard ISO 7726 - 1998.

Air samples were collected at 140 L/minute and at one meter tall using malt extract agar with chloramphenicol as a bacteria growth inhibitor (MEA), in the facilities, and also, outside premises, since this is the place regarded as reference. Swabs were performed according to the International Standard ISO 18593 - 2004, using a 10 cm square of metal disinfected with 70% alcohol solution between samples. Swabs were inoculated in triplicate on MEA and in mycobiotic agar with cycloheximide (MA). Subsequently, these were incubated at 27ºC (MEA for 5 to 7 days and MA for 15 to 20 days).

After laboratory processing and incubation of the collected samples, quantitative (CFU/m^3 and CFU/m^2) and qualitative results were obtained, with identification of isolated fungal species. Whenever possible, filamentous fungi were identified to the species level, since adverse health effects vary according to fungal species [10, 11]. Identification of filamentous fungi was carried out on material mounted in lactophenol blue and achieved through morphological characteristics listed in illustrated literature [11].

With the obtained data, tables with frequency distribution of isolated fungal species were made. Fungal concentration dependence in the two monitored environmental parameters – temperature and relative humidity – was also analyzed.

3 Results

Twenty different species of fungi in air were identified, being the 4 most commonly isolated the following genera: *Cladosporium* (40,5%), *Alternaria* (10,8%), *Chrysosporium* and *Aspergillus* (6,8%). Among *Aspergillus* genus, were identified the species *Aspergillus flavus, Aspergillus niger* and *Aspergillus fumigatus.* In addition to these genera, were also identified: *Fusarium* sp., *Fusarium incarnatum, Fusarium oxysporum, Exophiala werneckii, Stemphylium* sp., *Exophiala* sp., *Phoma* sp., *Scytalidium* sp., *Aureobasidium* sp., *Mucor* sp., *Penicillum* sp., *Ulocladium* sp. and *Rhizopus* sp.

Concerning surfaces, 21 different species of fungi were identified, being the 4 genera more identified *Penicillium* sp. (51,8%), *Cladosporium* sp. (25,4%), *Alternaria* sp. (6,1%) and *Aspergillus* sp. (4,2%). Among *Aspergillus* genus, were identified the species *Aspergillus glaucus*, *Aspergillus. fumigatus* and *Aspergillus niger.* In addition to these genera, were also identified: *Cladosporium sphaerosperma*, *Chrysosporium* sp., *Trichothecium roseum*, *Graphium* sp., *Scopulariopsis* sp., *Fusarium oxysporum, Trichoderma* sp., *Exophiala* sp., *Chrysonilia* sp., *Scytalidium* sp., *Gliocladium sp., Ulocladium* sp., *Mucor* sp. and *Scedosporium prolificans* sp. (Table 1).

Regarding comparison of concentrations found in air, for indoor and exterior environments, all indoor areas showed less contamination than exterior areas. However, all the indoor spaces presented fungal species different from the ones isolated outdoor. Some fungi that were only isolated indoor were: *Aspergillus flavus, Aspergillus niger, Aspergillus fumigatus, Phoma* sp., *Aureobasidium* sp., *Mucor* sp., *Fusarium clamidosporos* sp., *Fusarium incarnatum, Fusarium oxysporum* and *Rhizopus* sp..

Outside premises *Cladosporium*, *Alternaria* and *Chrysosporium* were the prevailing genera.

Concerning quantitative results the highest fungal contamination found indoor was 240 CFU/m^3, and outside premises was 740 CFU/m^3.

Table 1: Most frequent fungi identified in the poultry air and surfaces.

Air	Frequency (%)
Cladosporium sp.	40,5
Alternaria sp.	10,8
Chrysosporium sp.	6,8
Aspergillus sp.	6,8
Mucor sp.	2,7
Penicillium sp.	2,7
Scytalidium sp.	2,7
Others	27
Surfaces	**Frequency (%)**
Penicillium sp.	51,8
Cladosporium sp.	25,4
Alternaria sp.	6,1
Aspergillus sp.	4,2
Chrysosporium sp.	2,9
Others	9,6

Regarding the influence of environmental variables monitored no significant correlation ($p > 0,05$) was reveal. Temperature and relative humidity contributed only in 2,75% and 59,0 %, respectively, to CFU/m^3 variation explanation (Figure 1 and Figure 2).

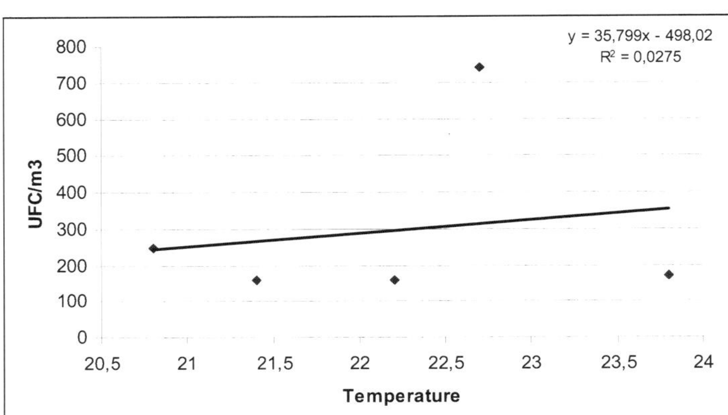

Figure 1: Influence of temperature in CFU/m^2.

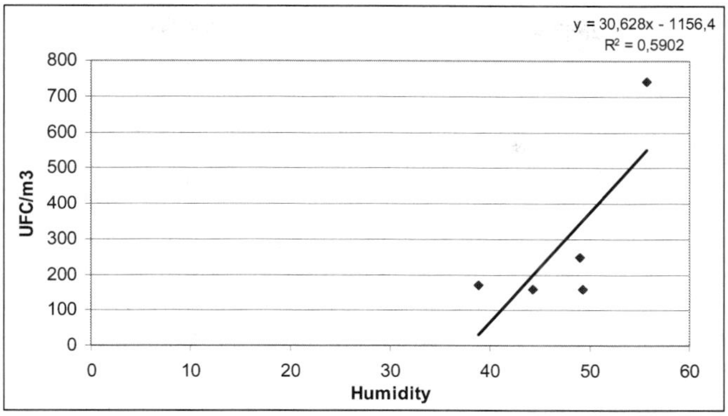

Figure 2: Influence of relative humidity in CFU/m^2.

4 Discussion

Cladosporium genus, predominant type in the air, is probably the fungus that occurs more frequently around world, especially in temperate climates [12]. The same genus is deeply connected to indoor condensation problems [13]. For *Penicillium* sp., predominant type in the surfaces, there are different potential risks associated with their inhalation, due to the toxins release. In *Alternaria* sp. case, second in air and third in surfaces, there are potentially allergic effects, only because the spores presence [14].

With regard to qualitative assessment of fungal contamination in air, it is suggested that, among other species, *Aspergillus fumigatus* and *Penicillium, Trichoderma*, *Fusarium* and *Ulocladium* species, all of them isolated in the present study, are regarded as indicators of humidity problems or potential risk to health [1]. Moreover, according to American Industrial Hygiene Association (AIHA), in 1996, for determination of biological contamination in environmental samples, confirmed presence of the species *Aspergillus flavus* and *Aspergillus fumigatus*, both identified in this study, requires implementation of corrective measures [15]. Also noteworthy is the fact that *Aspergillus fumigatus*, isolated in air and surfaces, is one of the saprophytic fungi most widespread in air and is capable of causing severe or sometimes fatal aspergillosis [16].

In addition, *Aspergillus flavus*, also isolated in the poultry air, is a well-known producer of potent mycotoxins (aflatoxins) and besides this species other isolated genera, like *Fusarium* and *Penicillium,* are also known as mycotoxins producers [15].

For quantitative assessment in air (CFU/m^3) is proposed corrective measures implementation whenever, in a given space, one or more of the following conditions were verified: a) > 50 CFU/m^3 of a single fungal species; b) > 150 CFU/m^3 if several fungal species are isolated; c) > 300 CFU/m^3 if there are mainly filamentous fungi [17]. The first condition a) was found for *Cladosporium* sp. and *Exophiala* sp. and the second condition b) was found in all the interior spaces monitored. However, and because fungal concentration indoor was considerably lower than one outdoor, some authors consider that there should be no concerns about a possible fungal contamination [14]. Taking into account what is mentioned in Portuguese law, 500 CFU/m^3 maximum reference concentration in interior air, was not exceeded in any of the monitored spaces. In one interior space, where 240 CFU/m^3 were isolated, revealed the highest fungal contamination.

Also worth mentioning is the fact that outdoor air is a major source of fungi in indoor, thus justifying the coincidence between prevailing genera, *Cladosporium*, *Alternaria* and *Chrysosporium*, in both these environments [18]. Nonetheless, all monitored interior spaces had fungal species different from the ones isolated outside, suggesting fungal contamination from within [19].

Surfaces sampling, in addition to air sampling, is essential to achieve the fungal contamination characterization and evaluation, and can be used to identify contamination sources [20, 21]. There was some coincidence between species isolated in air and on surfaces and, furthermore, should be taken into account that species present only on surfaces may aerosolized and become airborne, depending on activities carried out [22] and its occupants [23], whereas in this case will be not only farmers, but also chickens.

Taking into account the isolated fungal species in poultry air and surfaces, we have to consider not only the occupational health problem, due to the *Aspergillus fumigatus* presence, among others species, but also the public health problem since there were some fungal species that release mycotoxins [15], and there is evidence that mycotoxins can cause human disease from inhalation and from ingestion of fungus-contaminated food [9]. Besides that, early studies did provide data illustrating that increasing human hepatocellular carcinoma rates corresponded to increasing levels of dietary aflatoxins exposure [24]. The requirements for aflatoxin production are relatively non-specific, since moulds can produce them on almost any foodstuff, with therefore a wide range of commodities contaminated at final concentrations which can vary from < 1 μg/kg (1 p.p.b.) to 12 000 μg/kg (12 p.p.m) [25]. Because measurement of human exposure to aflatoxin, by sampling foodstuffs or by dietary questionnaires is extremely imprecise, it's important to consider aflatoxin exposure biomarkers since they have great potential for accurate assessment of exposure [26].

Results related to environmental variables are not consistent with the expected [27]. It was found that the relationship between the fungal air contamination and the temperature and relative humidity was not statistically significant (p>0,05). This may be justified by the effect of other environmental variables also influencing fungal spreading, namely workers and chickens, who may carry, in their own body (commensal flora) or clothing, a great diversity of fungal species [23], as well the developed activities that may also affect fungal concentration [22].

5 Conclusions

It was possible to characterize fungal distribution in poultry air and surfaces, evaluate the association of environmental variables with this distribution and recognize the public heath problem, because there is scientific evidence that mycotoxins can cause human disease after inhalation or ingestion of fungus-contaminated food. Furthermore, also worth mentioning is the occupational health problem due to the presence of *Aspergillus fumigatus* in poultry air and surfaces.

Unlike other studies, environmental variables monitored (temperature and relative humidity) did not show the expected association with fungal concentration, which may possibly have resulted from other variables not investigated in this study.

References

[1] Goyer N, Lavoie J, Lazure L & Marchand G., *Bioaerosols in the Workplace: Evaluation, Control and Prevention Guide.* Institut de Recherche en Santé et en Sécurité du Travail du Québec, 2001.

[2] Daisey J, Angell W & APTE M., Indoor air quality, ventilation and health symptoms in schools: an analysis of existing information. *Indoor Air,* 13, pp., 53 – 64, 2003.

[3] Jarvis B., *Stachybotrys chartarum*: a fungus of our time. *Phytochemistry,* 64, pp. 53 – 60, 2003.

[4] Molocznik A., Qualitative and quantitative analysis of agricultural dust in working environment. *Ann. Agric. Environm. Med.,* 9, pp., 71 – 78, 2002.

[5] Buys E, Nortjé G, Jooste P & Von Holy A., Bacterial population associated with bulk-packaged beef supplemented with dietary vitamin E. *Int. J. Food. Microbiol*; 50, pp., 239 – 244, 2000.

[6] Borch E & Arinder P., Bacteriological safety issues in red meat and ready-to-eat meat products, as well as control measures. *Meat Sci.*, 62, pp., 381 – 390, 2002.

[7] White P, Collins J, McGill K, Monahan C & O'Mahony H., Distribution and prevalence of airborne microorganisms in three commercial poultry processing plants. *J. Food Products*, 64, pp., 388 – 391, 2001.

[8] Lues J, Theron M, Venter P & Rasephei M., Microbial composition in bioaerosols of a high-throughput chicken-slaughtering facility. *Poultry Science*, 86, pp., 142 – 149, 2007.

[9] Shlosberg A, Zadikov I, Perl S, Yakobson B, Varod Y & Elad D., *Aspergillus clavatus* as the probable cause of a lethal mass neurotoxicosis in sheep. *Mycopathology*, 114 (1), pp., 35 – 9, 1991.

[10] Rao C, Burge H & Chang J., Review of quantitative standards and guidelines for fungi in indoor air. *J Air Waste Manage Assoc.*, 46, pp. 899 – 908, 1996.

[11] Hoog C, Guarro J, Gené G & Figueiras M., (2th ed). *Atlas of Clinical Fungi.* Centraalbureau voor Schimmelcultures, 2000.

[12] Cooley J, Wong W, Jumper C & Straus D., Correlation between the prevalence of certain fungi and sick building syndrome. *Occup. Environ Med*, 55, pp., 579 – 584, 1998.
[13] Garret M, Rayment P, Hooper M, Abranson M & Hooper B., Indoor Airborne Fungal Spores, House Dampness and Associations with Environmental Factors and Respiratory Health in Children. *Clinical and Experimental Allergy,* 28, pp., 459 – 467, 1998.
[14] Kemp P, Neumeister-Kemp H, Esposito B, Lysek G & Murray F., Changes in airborne fungi from the outdoors to indoor air; Large HVAC systems in nonproblem buildings in two different climates. *American Industrial Hygiene Association,* 64, pp. 269 – 275, 2003.
[15] American Industrial Hygiene Association: *Field Guide for the Determination of Biological Contaminants in Environmental Samples.* AIHA, 1996.
[16] Yao M & Mainelis G., Analysis of Portable Impactor Performance for Enumeration of Viable Bioaerosols. *Journal of Occupational and Environmental Hygiene*, 4, pp., 514 – 524, 2007.
[17] Miller J., Fungi as Contaminants of Indoor Air. *Atmos. Environm,* 26 A, pp., 2163 – 2172, 1992.
[18] Nevalainen A., Bio-aerosols as exposure agents in indoor environment in relation to asthma and allergy. *Section 3 Asthma and allergy. Proceedings of the First ENVIE Conference on Indoor Air Quality and Health for EU Policy, Helsinki, Finland, 2007.*
[19] Kemp P, Neumeister-Kemp H, Murray F & Lysek G., Airborne fungi in non-problem buildings in a southern-hemisphere Mediterranean climate: preliminary study of natural and mechanical ventilation. *Indoor and Built Environment*, 11; pp., 44 – 53, 2002.
[20] Stetzenbach L, Buttner M & Cruz P., Detection and enumeration of airborne biocontaminants. *Current Opinion in Biotechnology*, 15, pp. 170 – 174, 2004.
[21] Klánová K & Hollerová J., Hospital indoor environment: screening for micro-organisms and particulate matter. *Indoor and Built Environment*, 12, pp. 61 – 67, 2003.
[22] Buttner M & Stetzenbach L., Monitoring Airborne fungal spores in an experimental indoor environment to evaluate sampling methods and the effects of human activity on air sampling. *Applied and Environmental Microbiology*, 59, pp. 219-226, 1993.
[23] Scheff P, Pulius V, Curtis L & Conroy L., Indoor air quality in a middle school, Part II: Development of emission factors for particulate matter and bioaerosols. *Applied Occupational and Environmental Hygiene*, 15, pp. 835 – 842, 2000.
[24] Bosch F & Munoz N., Prospects for epidemiological studies on hepatocellular cancer as a model for assessing viral and chemical interactions. *IARC Scientific Publications*, 89, pp. 427 – 438, 1988.

[25] Ellis W, Smith J, Simpson B & *et al.,* Aflatoxins in food – occurrence, biosynthesis, effects on organisms, detection, and methods of control. *Crit Rev Food Sci Nutr*, 30, pp. 403 – 439, 1991.
[26] Wild C & Turner P., The toxicology of aflatoxins as a basis for public health decisions. *Mutagenesis*, 17, pp. 471 – 481, 2002.
[27] Kakde U, Kakde H & Saoji A., Seasonal Variation of Fungal Propagules in a Fruit Market Environment, Nagpur (India). *Aerobiologia*, 17, pp. 177 – 182, 2001.

Factors controlling the release of arsenic from mining tailings

B. E. Rubio-Campos[1], I. Cano-Aguilera[1],
A. F. Aguilera-Alvarado[1], G. De la Rosa[1] & S. H. Soriano-Pérez[2]
[1]*Department of Chemical Engineering, University of Guanajuato, Mexico*
[2]*Department of Chemical Sciences,*
Autonomous University of San Luis Potosi, Mexico

Abstract

Some mine tailings pools in the Mine District of Guanajuato, Mexico, present a varied distribution and temporal and spatial concentration of elements that are potentially toxic, such as manganese, cadmium and zinc. These elements were detected in majority concentrations, and arsenic was present in the two major oxidation states As(III) and As(V). The highest arsenic concentration in the surrounding surface water reservoirs was detected when a rainy seasons occurred, which in turn is mainly a function of pH and the presence of bicarbonate ions. The conceptual model to describe the mobilization of arsenic from mining tailings towards the aqueous systems proposes a scenario where oxidation, the neutralization of acid drainage by carbonates, and arsenic desorption by bicarbonates takes place in different steps and at different times.
Keywords: mining tailings, potentially toxic elements, arsenic release.

1 Introduction

The mining district of Guanajuato is located 475 km from Mexico City. It is considered one of the largest worldwide. However, large amounts of mining tailings, which result from crushing and milling ore, once they have been recovered through commercial metal physical or chemical processes, have been generated over time [1]. These can be transported and become a severe environmental problem in relation to soil, sediment, surface water and groundwater pollution [2].

www.witpress.com, ISSN 1743-3541 (on-line)
doi:10.2495/ETOX100061

Mining Tailings of *San Nicolas* Mine (*JSN*) and mining tailings of *Peregrina* Mine (*JP*) are both located in this district and are already abandoned. Under certain physico-chemical, geological and biological conditions they can promote the leaching and transportation of metals to aqueous systems. Several studies have been carried out aimed at different purposes: the identification of potentially toxic elements in leaching experiments of mining tailings [3]; geological and mineralogical characterization of mining tailings [4]; and the arsenic potential release from mining tailings [5]. In this context and in order to explain the presence of arsenic and other toxic elements in water from *Presa de Mata*, it is important to locate and characterize the source of the pollution and the factors that are controlling this event.

2 Methodology

2.1 Sampling and characterization of mining tailings in this study

2.1.1 Sampling, preservation and transportation of mining tailings in this study

Mining tailing samples of 2 k were collected from the surface and from 10 cm deep [6]. The choice of the place and sampling sites carried out depended on local conditions and ease of access to sampling points. Once in the laboratory all samples were air dried and sieved to homogenize them. Sampling was performed at three different seasonal times, corresponding to an abundant precipitation season (August 2007), a dry cold season (February 2008) and a dry season (July, 2009). The sampling of mining tailings of *Peregrina* mine was carried out only in the dry season (July, 2009).

2.1.2 Physico-chemical analysis of samples of mining tailings in this study

Granulometric analysis was performed. Subsequently, the physico-chemical properties – humidity percent, real density, bulk density, porosity and pH – were measured. For the determination of arsenic (As), cadmium (Cd), lead (Pb), manganese (Mn), and zinc (Zn), the mining tailing samples were digested in a microwave oven (Perkin Elmer, Multiware 3000) [7]. Element analyses were determined by flame atomic absorption spectroscopy (FAAS) (Perkin Elmer, AAnalyst 100). The hydride generation technique was coupling to FAAS for arsenic determination.

A speciation spectrum of arsenic by X-ray Absorption Spectroscopy was obtained. The collected samples of *JSN* packaged subsequently were analyzed beamline 7-3 at the Stanford Synchrotron Radiation Laboratory (SSRL) in Menlo Park, CA.

2.1.3 Mineralogical analyses of samples of mining tailings in this study

Four samples of 10 g of mining tailing samples in this study were dried and sieved (*JSN* and *JP*); later the samples were concentrated using a solution of sodium dodecyl sulfate (SDS) in order to eliminate the high silica content. Two of these samples' mineralogical composition were obtained using an X-ray diffractometer (Siemens D-500) with a copper anode (λ=1.5418 Å) operated at 30 kV and 20 mA. Another two samples were determined by elemental qualitative analyses by X-ray fluorescence, using an X-ray spectrometer (Siemens SRS-200) operated at 40 kV and 30 mV.

2.1.4 Acid-base balance of samples of mining tailings in this study

The potential of acidity (PA) of the mining tailings was obtained, quantifying sulfides as the difference between total sulphur and sulfates. The potential of neutralization (PN) was obtained by the reaction of the mining tailings with sequential additions of HCl and pH after 2, 22 and 24 h. Finally the net potential of neutralization (PNN) was calculated by:

$$PNN = \frac{PN}{PA} \tag{1}$$

When the relationship (1) is less than 1.2 the tailings are generators of acid rock drainage, otherwise they do not generate acid rock drainage.

2.2 Sampling and characterization of aqueous samples of the main tributaries to *Presa de Mata*

2.2.1 Sampling, preservation and transport of potentially toxic elements from aqueous samples of the main tributaries to *Presa de Mata*

Collection containers (1 l approximately) were rinsed two or three times with the liquid to take as a sample [8]. Aqueous samples were acidified with nitric acid until a pH of less than or equal to 2.0 was obtained for metal analyses. The aqueous samples for anions determination were refrigerated until the analyses (sulfates, chlorides, carbonates-bicarbonates).

2.2.2 Physico-chemical analyses of potentially toxic elements in aqueous samples from the main tributaries to *Presa de Mata*

The pH and temperature of water samples *in situ* were measured using a Corning Checkmate II Modular Meter System, in accordance with the instructions of the manufacturer. Element concentration of water samples were determined by FAAS. The hydride generation technique was coupled for each arsenic determination. The following anions, sulfates, chlorides, carbonates and bicarbonates were also determined.

2.3 Leaching of potentially toxic elements of mining tailings in this study

2.3.1 Leaching of potentially toxic elements of mining tailings in a batch system

Samples of *JP* were used and contacted with different leachate solutions, in order to simulate different conditions that may be carried out. The leachate solutions employed were: acidified water pH 4, 9K culture medium, and bacteria growing in 9K cultured medium [9]; the final concentration of each array was 20 mg mining tailings/ml solution. For each experiment, 40 ml of each array were placed in conical polypropylene tubes in triplicate. The tubes were incubated for 1, 3, 5, 7 and 14 days in agitation at room temperature. Metals in the solution were determined. The metals analyzed were: As, Cd, Mn, Ag, Pb and Zn.

2.3.2 Leaching of potentially toxic elements of mining tailings in a continuous system

Samples of mining tailings, *JSN* and *JP*, were dried at room temperature previously, sieved (< 0.5 mm) and packed in glass columns of 4 cm in diameter and 16 cm high. Nitric acid solution (pH 4 ± 0.1 – maximum 2 ppb of metal impurities) was introduced in an up stream flow using a peristaltic pump at 12 and 9 ml/min for packed columns with samples of *JSN* and *JP*, respectively. Five fractions of leachate of each column by triplicate were collected at different times [10]. Each fraction was collected over the top of the column. The parameters in the leachate: pH, oxide-reduction potential, electrical conductivity were registered. The concentration of elements was determined by FAAS and As was determined by hydride generation coupling to FAAS.

3 Results and discussion

3.1 Sampling and characterization of mining tailing samples in this study

3.1.1 Sampling, preservation and transportation of mining tailing samples in this study

A map of microwatershed *Peregrina-Presa de Mata-Monte de San Nicolas* is displayed in fig. 1. This was created with Arcview GIS version 3.2 and shows

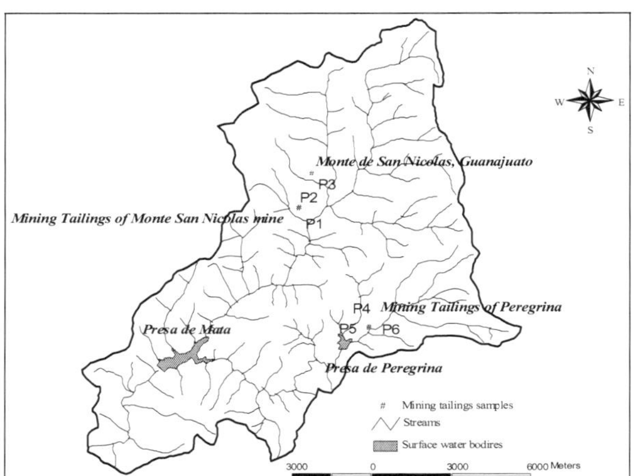

Figure 1: Microwatershed *Monte de San Nicolas-Presa de Mata-Peregrina*, Guanajuato. Mining tailing sampling points are located in (P) places.

the main surface water bodies towards *Presa de Mata*, as well as the points of the mining tailing samples. Georeferenced points: P1, P2, P3 and P4, represent sampling sites of *JSN* and P5 and P6 represent sampling sites of *JP*.

3.1.2 Physico-chemical analysis of mining tailing samples in this study

The physico-chemical parameters of the mining tailing samples determined are displayed in table 1.

Samples of *JP* showed a fine particle size. This feature is important for oxidation processes because it allows better oxygen diffusion, water conduction and interaction with other reactive phases.

Element concentrations in *JSN* and *JP* and their respective standard deviations are shown in tables 2 and 3, respectively.

Table 1: Physico-chemical parameters determined in mining tailing samples.

Sample (P)	Real density (g/cm^3)	Bulk density (g/cm^3)	Porosity	Humidity (%)	pH
1	2.97	1.01	0.79	0.274	7.60
2	2.86	1.38	0.52	0.102	7.21
3	3.22	1.01	0.68	0.274	6.08
4	2.87	0.92	0.67	0.288	7.08
5	2.22	1.05	0.52	0.110	6.00
6	2.20	1.03	0.50	0.103	6.03

Table 2: Concentration of elements in mining tailings *JSN*.

Sampling season	As (mg/k)	Mn (mg /k)	Zn (mg/k)	Cd (mg/k)	Pb (mg/k)
August 2007	4.8±0.004	638.9±0.025	219.11±3.25	No detected	34.82±0.04
February 2008	21.7±0.020	1030.1±0.001	266.77±3.1	2.58±0.006	63.85±0.11
July 2009	11.1±0.011	1960.3±0.001	278.52±1.2	3.33±0.003	101.8±0.005

Table 3: Concentration of elements in mining tailings *JP*.

Sampling season	Pb (mg/k)	As (mg/k)	Mn (mg/k)	Zn (mg/k)	Cd (mg/k)
February 2008	86.99±0.15	14.0±0.005	1388.2±0.001	484.19±0.9	2.21±0.004
July 2009	75.04±0.25	17.22±0.017	1165.3±0.001	379.12±1.5	2.57±0.009

The total As contents ranged between 4.8 and 11.1 mg/k. This concentration does not exceed the permissible maximum limit (*LMP*) for As, which corresponds to 22 mg/k [11], but the potential risk is presented because there are variations of arsenic concentrations over time and space. Concentrations of Mn show diversity in time and space. The Zn contents in all concentrations exceed the Canadian Standards [12]. The Pb contents ranged between 34.8 and 101.8 mg/k. In rainy seasons the presence of Cd was not detected. The concentration of elements in *JSN* occurs in the following sequence: Mn > Zn > Pb > As > Cd.

The major elements in *JP* were Mn and Zn. The As contents ranged between 14 and 17.22 mg/k, showing a greater concentration in the dry season. Therefore, the concentration of elements in *JSN* presents the same sequence of that in samples of the *Monte de San Nicolas* mine, but in a higher proportion.

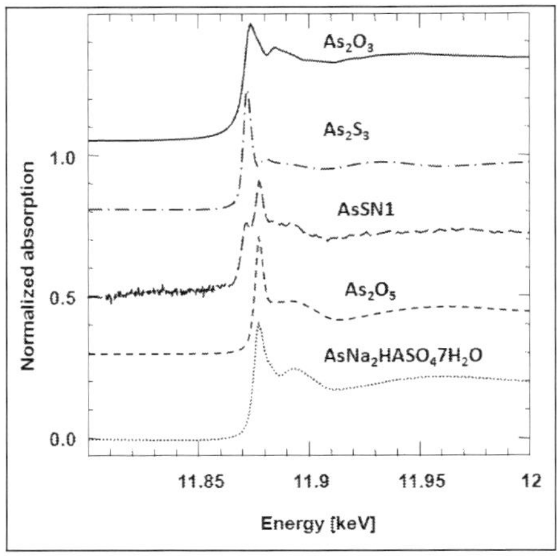

Figure 2: Spectra of X-ray absorption for sample of *JSN* for As (*AsSN1*). Model compounds: As_2O_3, As_2S_3, As_2O_5 y $AsNa_2HAsO_4\,7H_2O$.

Table 4: Mineralogical analyses of mining tailings samples in this study by X-ray diffraction.

JSN		*JP*	
Phases identified	Possible phases	Identified phases	Possible phases
Quartz SiO_2 Calcite $CaCO_3$ Plagioclase $NaAlSi_3O_8$ Clinochlore $Mg_5Al(Si_3Al)O_{10}(OH)_8$ Muscovite $KAl_2Si_3AlO_4(OH)_2$	Smectite Fe_2O_3	Quartz SiO_2 Calcite $CaCO_3$ Ortoclase $KAlSi_3O_8$ Plagioclase $NaAlSi_3O_8$ Smectite (montmorillonite) Kaolinita $Al_2Si_2O_5(OH)_4$	Hematite Fe_2O_3

Spectra of several model compounds for As and *JSN* are shown in fig. 2. This figure shows displacement in absorption energy for As(V) and As(III). The As from mining tailing samples (*AsSN1* in the spectrum) displays two absorption energies for the same element. The first line threshold occurs at 11,867 keV and the second at 11,875 keV. This corresponds to the presence of As(III) and As(V), respectively.

3.1.3 Mineralogical analyses of mining tailing samples from *Monte San Nicolas* and *Peregrina* mines

The results of the X-ray diffraction are displayed in table 4. This table shows that the materials present a typical mineralogical composition of mining tailings. It was not possible to identify the presence of metal sulfides, given such high levels in the content and crystallization of the quartz and calcite minerals present.

Elemental qualitative analyses by X-ray fluorescence show that *JSN* and *JP* are constituted by Fe, Ca, Sr, Zn and Rb. Both samples contain Fe as the majority element, which is typical of mineral composition; also these samples consist of calcium, which reflected the high content of carbonates and that contribute to neutralize the degree of acidity, reducing the acid rock drainage.

3.1.4 Acid-base balance of mining tailings samples in this study

The determination of the acid rock drainage of tailings due to their reactivity can be determined by the oxidation of sulfide. The results are presented in table 5. These results suggest that the mining tailings in this study are not acid generators of acid rock drainage; this is possibly due to the high amount of carbonates containing mining waste and that they somehow contribute to neutralize the degree of acidity that is generated or generated since the age of abandonment of the mining tailings and processes that could have been carried out.

Table 5: Determination of acid rock drainage in the mining tailings.

Sample	PN (Kg $CaCO_3$/ton mining tailings)	PA (Kg $CaCO_3$/ton mining tailings)	(PNN)
P1	77.0	7.75	9.93
P2	76.0	7.7	9.87
P3	82.5	7.06	11.69
P4	80.8	7.02	11.42
P5	175.0	4.86	36.04
P6	185.0	4.13	44.75

3.2 Sampling and characterization of aqueous samples in the main tributaries to *Presa de Mata*

3.2.1 Sampling, preservation and transport of potentially toxic elements from aqueous samples in the main tributaries to *Presa de Mata*

A map of microwatershed *Presa de Mata, Guanajuato* and its main tributaries is presented in fig. 3, where sampling points are indicated for various seasonal times. The sampled surface water bodies are located "downstream" from deposits of mining tailings, which are considered as sources of pollution.

The results of potentially toxic elements corresponding to three seasons of sampling in the stream from *Monte de San Nicolas* to *Presa de Mata* showed that the As contents ranged between 0.012 and 0.015 mg/l, which is within Mexican Standards [13], but is not within the limit recommended by international guidelines, since chronic exposure to As causes toxic effects to human health. These results suggest that detected elements can be leachated from mining tailings and transported by main streams to *Presa de Mata*. In the case of the dry season and abundant rainfall, samples have higher concentrations of As. High concentrations of Mn are detected. Pb and Zn were only identified in the dry season (July, 2009) and rainy season (August, 2007), respectively. The Pb contents exceeds the *LMP*, while the Zn content is lower than this standard. Water samples from *Peregrina to Presa de Mata* corresponding to July 2009 presented high values of both As and Mn.

Variation of pH in the study sites for sampling in the dry season ranged between 7 and 8; this indicates a neutral environment to slightly alkaline, which can be explained by the presence of carbonates in samples of tailings that neutralize the degree of acidity that could occur. The temperature registered ranged between 19 and 24 °C.

Concentrations of anions were determined by following established guidelines in the Mexican Standards. The results are displayed in table 6, which

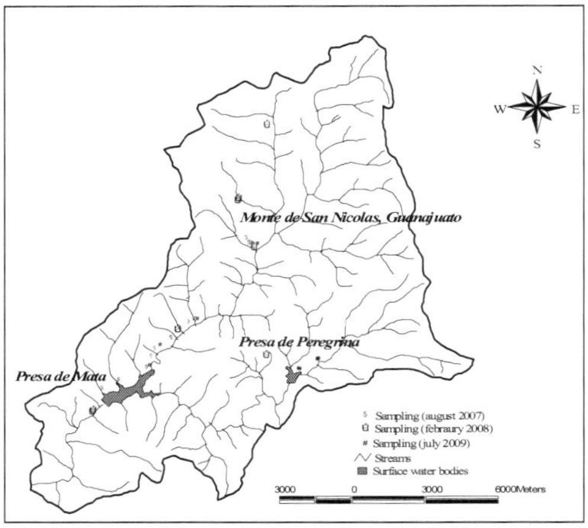

Figure 3: Microwatershed *Monte de San Nicolas-Presa de Mata-Peregrina, Guanajuato*. Sampling points are shaded in gray.

Table 6: Anions in water samples (July, 2009).

	Sample	Cl^- (mg/l)	HCO_3^- (mg/l)	SO_4^{2-} (mg/l)
Stream from *Monte San Nicolas* to *Presa de Mata*	1	38.67±7.4	162.7±3.7	85.1±1.8
	2	38.67±3.7	162.7±3.7	128.6±1.5
	3	32.23±3.7	162.7±3.7	98.2±1.4
	4	32.23±7.4	122.0±3.7	311.4±1.3
	5	38.67±7.4	122.0±3.7	387.5±1.9
Stream from *Peregrina* to *Presa de Mata*	6	32.23±3.7	122.0±0.005	447.3±1.8
	7	38.67±3.7	203.3±0.005	154.7±1.5
	8	38.67±3.7	203.3±0.005	190.5±1.4
	9	32.23±3.7	122.0±0.005	447.3±1.3
	10	38.67±3.7	203.3±0.005	190.5±1.9

correspond to streams from *Monte San Nicolas* and *Peregrina* mining tailing sites, respectively.

3.3 Leaching of potentially toxic elements of mining tailing samples from *Monte San Nicolas* and *Peregrina* mines

3.3.1 Leaching of potentially toxic elements of mining tailings from *Peregrina* mine in the batch system

Fig. 4 shows the concentration of As leachate depending on the time, for different leachate solutions. In this figure, As was leachated in low amounts when a 9K medium and bacteria type *Thiobacillus ferrooxidans,* previously selected from mining tailings and grown in a 9K culture medium, was used as a leachate solution. The continuous line represents the initial concentration of total As in the sample of mining tailings.

The leaching kinetic of Cd is displayed in fig. 5. Cd was leachated efficiently, since more than 90% of mining tailings were detected in the solution. The continuous line represents the initial concentration of the total Cd in the sample of mining tailings.

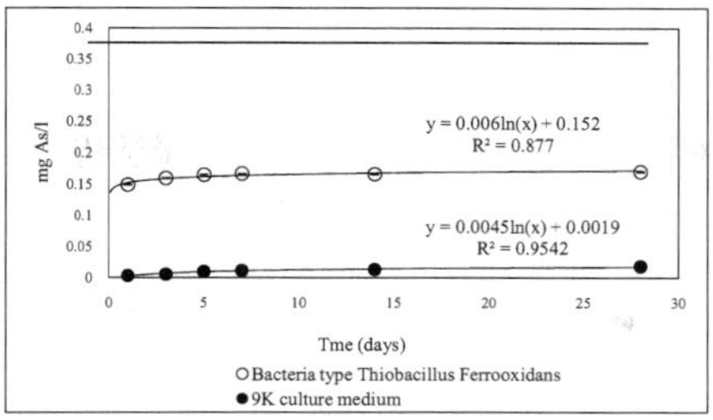

Figure 4: As leaching mining tailings in batch system.

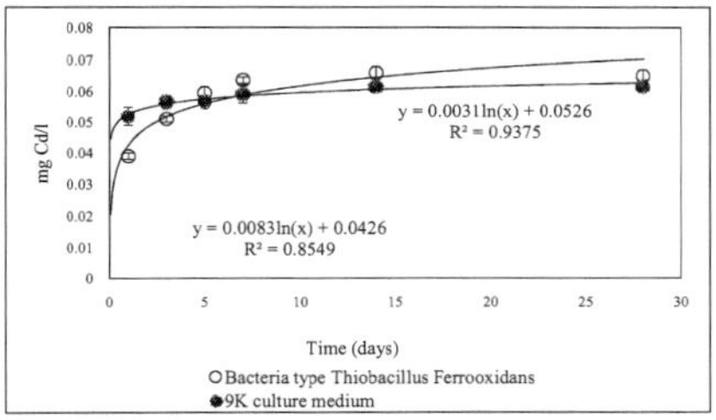

Figure 5: Cd leaching from mining tailings in the batch system.

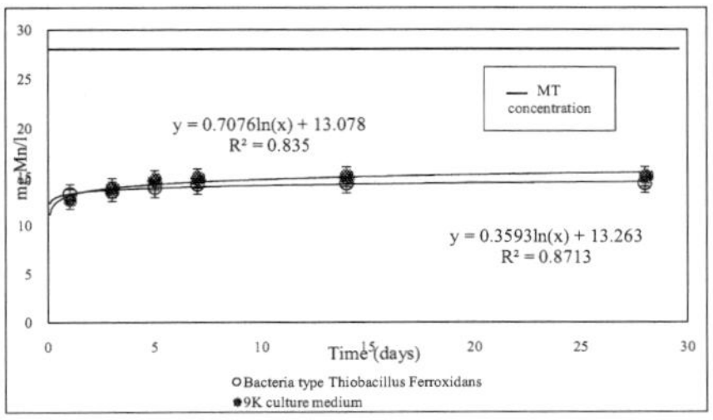

Figure 6: Mn leaching from mining tailings in the batch system.

For Mn, the leachate solution with bacteria was fitted to a logarithmic function as shown in fig. 6, where the trend line represents this behavior and points to the experimental data. The continuous line represents the initial concentration of total Mn in the sample of mining tailings.

The Pb leaching kinetic using 9K medium was fitted to a logarithmic function, while for Zn with bacteria type *Thiobacillus ferrooxidans* it was fitted to a linear function (these data are not shown). These results suggest that there are special sceneries that promote the leaching of potentially toxic elements and when these are combined in the natural system, their concentrations are higher in aqueous systems.

3.3.2 Leaching of toxic potentially elements of mining tailing samples in this study

The kinetics of As leaching in continuous systems as a function of time are displayed in fig. 7.

Each point represents the As released from mining tailings estimated by the difference between the total As concentration in the mining tailings and the As concentration detected in solution. In both cases the observed trend is logarithmic.

The variation of Mn concentration in the leachated solution and the fitted functions are shown in fig. 8. This figure shows that the concentration of Mn released from *JSN* is less than Mn from *JP*.

The Zn content in the solution and the fitted function are displayed in fig. 9. This element was obtained in higher concentration in comparison with other potentially toxic elements detected in the same system.

4 Conclusions

Mining tailing deposits from *Monte San Nicolas* and *Peregrina* showed a wide distribution and temporal and spatial concentration of potentially toxic elements, such as Mn, Cd and Zn in majority concentrations, as well as As. Arsenic is present in the two oxidation states (III) and (V).

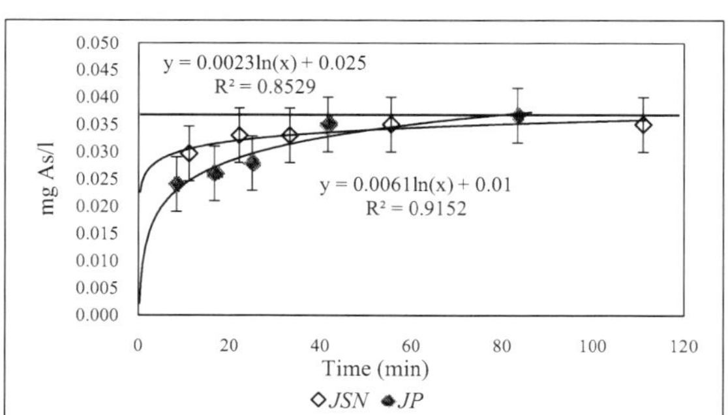

Figure 7: As leaching from mining tailings in a continuous system.

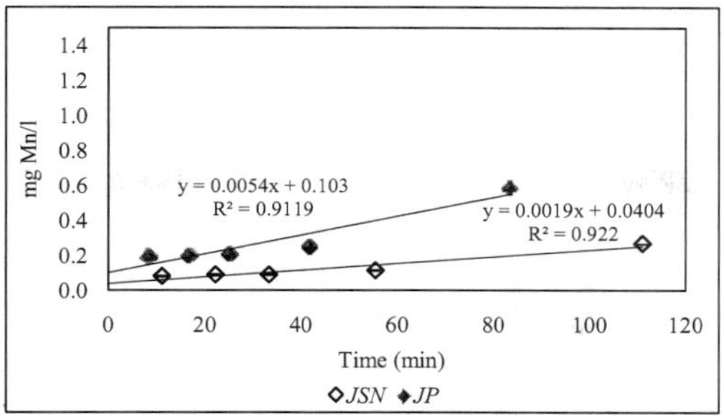

Figure 8: Mn leaching from mining tailings in the continuous system.

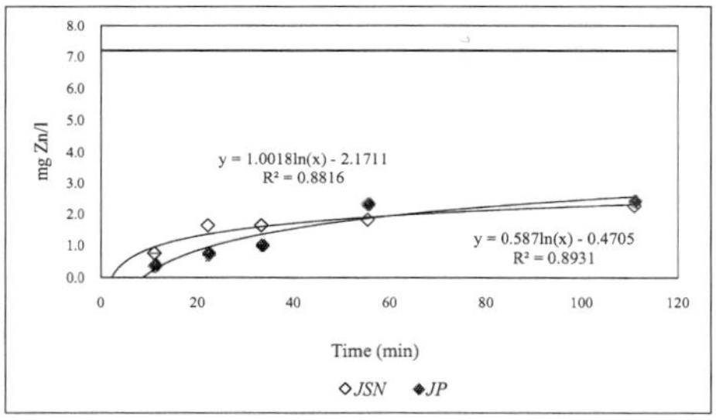

Figure 9: Zn leaching from mining tailings in the continuous system.

During rainy seasons, arsenic was obtained in higher concentration as a product of leaching of this element, which is a function of pH and the presence of bicarbonates in the system.

The pH of mining tailings in this study ranged between 6 and 8, which demonstrated that mining waste possesses a high amount of carbonates that reduce the degree of acidity that could generate the due porosity and granulometry of mining tailings.

Microbial activity can leach considerably potentially toxic elements, mainly those such as Cd.

The kinetics of leaching obtained in batch tests showed linear and logarithmic tendencies for Zn and As in continuous tests, respectively.

References

[1] Medel, A., Ramos, S., Avelar, F. J., Godínez, L. A., & Rodríguez, F., Caracterización de Jales Mineros y evaluación de su peligrosidad con base en su potencial de lixiviación. *Revistas Científicas de América Latina y el Caribe, España y Portugal.* **35**, pp. 33-35, 2008.

[2] Armienta, M. A., Villaseñor, G., Rodríguez, R. & Mango, H. The role of arsenic-bearing rocks in groundwater pollution at Zimapán Valley, Mexico. *Environmental Geology*, **40(4, 5)**, pp. 571-581, 2001.

[3] Morton-Bermea, O., Carrillo-Chávez, A., Hernández, E. & González-Partida, E. Determination of Metals for Leaching Experiments of Mine Tailings: Evaluation of the Potential Environmental Hazard in the Guanajuato Mining District, Mexico. *Bulletin of Environmental Contamination & Toxicology*. **73(4)**, pp. 770–776, 2004.

[4] Ramos-Arroyo, Y. R., Siebe-Grabach, C. D. Características geológicas y mineralógicas e historia de extracción del Distrito de Guanajuato, México. Posibles Escenarios geoquímicos para los residuos mineros *Revista Mexicana de Ciencias Geológicas*. **21(2)**, pp. 268-284, 2004.

[5] Ramos-Arroyo, Y. R. & Siebe-Grabach, C. D. Estrategia para identificar jales con potencial de riesgo ambiental en un distrito minero: estudio del caso en el Distrito de Guanajuato, Mexico. *Revista Mexicana en Ciencias Geológicas*. **23(1)**, pp. 54-74, 2006.

[6] NOM-141-SEMARNAT-2003, Online. www.semarnat.gob.mx/leyesynormas/Normas%20Oficiales%20Mexicanas%20vigentes/NOM_141_SEMAR_03_13_SEP_04.pdf

[7] Microwave Assisted Acid Digestion of Sediments, Sludges, Soils, and Oils U. S. Environmental Protection Agency, 3051 method. Online. www.epa.gov/waste/hazard/testmethods/index.htm

[8] NMX-AA-014-1980, Receiver bodies sampling, Online. www.semarnat.gob.mx/leyesynormas/Normas%20Mexicanas%20vigentes/NMX-AA-014-1980.pdf

[9] Silverman, M. P., Lundgren, D. G., Studies on the chemoautothropic iron bacterium *Ferrobacillus ferrooxidans*: an improved medium and a harvesting procedure of securing high cells yields. *Journal of Bacteriology*. **77**, pp. 642-647, 1959.

[10] Netherlands Normalization Institute (1993b) NEN 7343, Leaching Characteristics of Building on Solid Waste Materials-Leaching Tests Determination of the Leaching of Inorganic Components from Granular Materials with the Column Test, The Netherlands.

[11] NOM-147-SEMARNAT/SSA1-2004, Que establece criterios para determinar las concentraciones de remediación de suelos contaminados por arsénico, bario, berilio, cadmio, cromo hexavalente, mercurio, níquel, plata, plomo, selenio, talio y/o vanadio. Online. www.semarnat.gob.mx/leyesynormas/normasoficialesmexicanasvigentes/Elaboracin%20conjunta%20%20con%20otras%20secretarias/NOM_147_SEMARNAT_SSA1_2004.pdf

[12] Canadian Soil Quality Guidelines for the Protection of Environmental and Human Health (1996), Online. www.intranet2.minem.gob.pe/web/archivos/dgaam/estudios/lazanja/Anexo_R.pdf

[13] NOM-127-SSA1-1994-2000, Norma Oficial Mexicana, Salud Ambiental agua para uso y consumo humano. Online. ww.salud.gob.mx/unidades/cdi/nom/127ssa14.html

Correlation between cluster analyses of *Salmonella* strains isolated from diarrhetic patients in Kuwait and biofilm formation

A. Al-Mousawi[1], A. Eissa[1], F. Abu-Zant[1], H. Drobiova[1], I. Al-Saif[2] & E. Al-Saleh[1]

[1]*Microbiology Program, Department of Biological Sciences, College of Science, Kuwait University, Kuwait*

[2]*Department of Microbiology, Food Laboratories, Public Health Laboratory, Ministry of Health, Kuwait*

Abstract

Salmonella is a highly diverse group of strains that belong to the *Enterobacteriaceae* and can cause many infections, such as diarrhea, pyrexia and septicemia, in humans and animals. One important virulence factor is the ability to form biofilm. In the present study, the potential to form biofilms by *Salmonella* strains isolated from diarrhetic patients was investigated and correlated with the strain type. Isolated bacteria were identified by sequencing of 16S rDNA. The potential of *Salmonella* to form biofilms was determined using bioluminescence microbial cell viability assay. In addition, the metabolic fingerprints of *Salmonella* were determined using the Biolog system, following the manufacturer's instructions. Cluster analysis based on catabolic activity and 16S rDNA of isolated strains showed the tendency of most Schwarzengrund (66.7% - 70%) and E5 strains (85%) to cluster individually, which implied the high distinctive genetic background of Schwarzengrund and E5 strains. On the other hand, Heidelberg and Paratyphi strains were clustered among other stains, which reflected the genetic resemblance of these strains to other *Salmonella* strains. The biofilm studies showed the high potential of the majority of E6 (60%) and Heidelberg (66.7%) strains to form biofilms, while low potential to form biofilms was displayed by 78% of Schwarzengrund strains.

Keywords: Salmonella, biofilm, 16S rDNA sequencing, metabolic fingerprint.

WIT Transactions on Ecology and the Environment, Vol 132, © 2010 WIT Press
www.witpress.com, ISSN 1743-3541 (on-line)
doi:10.2495/ETOX100071

1 Introduction

The enterobacteria are a heterogeneous group of Gram-negative rods that naturally inhabit the intestine of humans and animals. *Salmonella*, which is one of the important genera belonging to the *Enterobacteriaceae*, can cause many infections such as diarrhea, pyrexia and septicemia in humans and animals (D'Aoust [2]). *Salmonella* is usually found in poultry, poultry salads, meat, meat products, raw milk, shell eggs, egg custards, improperly cooked mayonnaise, ice cream, and sauces (Uyttendaele et al. [16]; Foley and Lynne [3]). Thus, *Salmonella* is one of the most common types of foodborne illnesses reported causing food poisoning or salmonellosis (D'Aoust [2]). Following the uptake of contaminated food or water, *Salmonellae* reach the intestine, cross the epithelial barrier, activate the host signal transduction cascades and induce the formation of membrane ruffles localized at the contact point between the bacterium and host cell (Jones et al. [5]; Vazquez-Torres et al. [17] and Rescigno et al. [12]). Ultimately *Salmonellae* are taken up in large vacuoles. The symptoms of salmonellosis develop within 12-36 hours of eating food containing *Salmonella*. Symptoms include nausea with vomiting, abdominal cramps and diarrhea, which can be severe (Tsolis et al. [8]; Kingsley et al. [8]; Ohl and Miller [11]). In recent years problems associated with *Salmonella* have increased considerably, both in terms of occurrence and severity of cases of human salmonellosis (Jewes [4]; Robertson et al. [14]).

An important factor influencing the pathogenicity of *Salmonella* is its ability to adhere to the host's intestinal surfaces prior to invasion (Robertson et al. [13]). This interaction is thought to depend upon bacterial-like adhesins recognizing specific glycoconjugate receptors on host cell surfaces. The possession of active flagella combined with chemotaxis is also an important factor in the pathogenicity of *Salmonellae* (Khoramian et al. [7]; Jones et al. [6]). Thus, variability in genotypic or phenotypic surface adhesion-related characteristics of different subspecies is expected to affect the establishment of *Salmonellae* populations in different hosts (Robertson et al. [13]) and thus its ability to cause outbreaks. One mechanism for ensuring survival in the host might be the differential biofilm-forming potential within a natural *Salmonellae* population. Biofilm formation is the net result of multiple interacting molecular events (Robertson et al. [13]) and is most conveniently measured at the phenotypic level. Thus, studies correlating strain types, antibiotic resistance and potential to form biofilm are required.

Many authorities require discrimination between different strains based on their metabolic pathways. Additionally, several high-resolution molecular fingerprinting techniques have been used to reveal species and subspecies diversity and provide tools to follow the persistence of particular infections, to recognize new infections and to assess the efficacy of control measures (Cheah et al. [1]). Typing methods based on comparisons of whole genomic DNA, plasmid DNA or specific genetic determinants have been used as supplementary techniques. Therefore, the aim of this work was to assess the ability of

Salmonella strains to form biofilm and the correlation of biofilm formation potential with the metabolic and the genetic background of the different strains.

2 Materials and methods

2.1 Sampling and initial identification of *Salmonella*

Salmonella strains were isolated from stool samples brought from diarrhetic patients to the public health laboratory, Ministry of Health. For the isolation of *Salmonella* from stools, stool samples were inoculated into selinite broth to enrich for *Salmonella*. After overnight incubation, loop-full from selinite broth were streaked onto Hektoen enteric agar and incubated at 37°C for 48 hours (Vernacchio et al. [18]). Suspected grown *Salmonella* colonies were inoculated into sterile Triple Sugar Iron (TSI) agar slants, incubated at 35°C for 24 hrs and *Salmonella* was identified following the Bacteriological Analytical Manual of the U.S. Food and Drug Administration (http://www.foodinfonet.com/publication/fdaBAM.htm). Identified *Salmonella* were stored in 15% glycerol nutrient broth at -40°C.

2.2 Molecular identification of *Salmonella* by 16S sequencing

Isolated bacteria were identified by sequencing of 16S rDNA. For this purpose, genomic DNA was purified from pure bacterial cultures using the Wizard Genomic DNA purification kit as recommended by the manufacturer (Promega). The concentration of extracted DNA were quantified by fluorometry with a model TK 100 fluorometer (Hoefer Scientific Instruments) by using the extended assay protocol of the manufacturer and then stored at -20°C. Then, 16S rDNA sequences were amplified from extracted DNA using 27F (AGAGTTTGATC(AC)TGGCTCAG) and 1492R (ACGG(CT)TACCTTGTTA CGACTT) primers (Kuske et al. [9]). All reactions were carried out in 25 μl volumes, containing 12.5 pmol of each primer, 200 μM of each deoxyribonucleoside triphosphate, 2.5 μl of 10x PCR buffer (100 mM Tris-HCl, 15 mM $MgCl_2$, 500 mM KCl; pH 8.3), and 0.5 U of *Taq* DNA polymerase (Applied Biosystems, UK), increased to 25 μl with sterile water. PCR was performed in a Thermocycler, GeneAmp (Applied Biosystems, UK) with the following thermocycling program: 5 minutes denaturation at 95°C, followed by 30 cycles of 1 minute denaturation at 95°C, 1 minute annealing at 55°C, 1 minute extension at 72°C, and a final extension step of 5 minutes at 72°C. PCR products were visualized by electrophoresis in 2% (wt/vol) agarose gels and with ethidium bromide (0.5 μg/ml) staining. Then, PCR products were used as templates for DNA sequencing reactions. The sequencing PCR conditions were the same as those described above. Amplified DNA was purified by using a QIAQUICK PCR cleanup kit (Qiagen, Inc.), and DNA concentrations were determined as mentioned previously. Approximately 100 ng of 16S rDNA will be used as a template in dye terminator cycle sequencing reactions (Applied Biosystems PRISM dye terminator cycle sequencing kit). The 16S rDNA sequences obtained were run against the data bases using the basic alignment search tool (BLAST) and assigned to recognized representatives of the main eubacterial lineages based on scores of 97% or higher.

2.3 Biofilm formation by *Salmonella*

The potential of *Salmonella* for forming biofilms was determined using bioluminescence microbial cell viability Assay (Promega). For this purpose, 100 µL of overnight bacterial cultures (0.5 McFarland standard) were transferred to sterile polystyrene 96 well microplates followed by another 100 µL of phosphate buffer (pH 7). Microplates were incubated at 37°C for 24 hrs. Then, plates were washed gently with sterile phosphate buffer; 100 µL of phosphate buffer was added followed by the addition of bioluminescence reagent. Contents of the plates were mixed on an orbital shake, incubated at room temperature for five minutes and luminescence values were taken on a microplate luminescence detector LD 400C (Beckman Coulter, USA). Control wells containing the phosphate buffer without cells were run to obtain a value for background luminescence.

2.4 Determination of metabolic fingerprinting of *Salmonella*

The metabolic fingerprints of *salmonella* were determined using a Biolog system following the manufacturer's instructions (OmniLog® ID System).

3 Results

Three different approaches were used to analyze and cluster isolated *Salmonella*. The phylogenetic analysis of the 16S sequences using 97% similarity index demonstrated the presence of five main phylotypes (Figure 1). Each phylotype was composed of different strain types. However, cluster analysis based on the ability of isolates to utilize different organic substrates showed the presence of highly diverse metabolic potentials of isolated *Salmonella* (Figure 2). Furthermore, isolated *Salmonella* were segregated into three main groups based on their potentials to form biofilm: isolates with high potential to form biofilm (Table 1(a)), isolates with low potential to form biofilm (Table 1(b)) and isolates unable to form biofilm (Table 1(c)). Each group contained different strain types.

4 Discussion

Cluster analysis based on the catabolic activity (Figure 2) and 16S rDNA (Figure 1) of isolated strains showed the tendency of most Schwarzengrund (66.7% - 70%) and E5 strains (85%) to clusters individually, which implied the high distinctive genetic background of Schwarzengrund and E5 strains. On the other hand, Heidelberg and Paratyphi strains were clustered among other strains, which reflected the genetic resemblance of these strains to other *Salmonella*

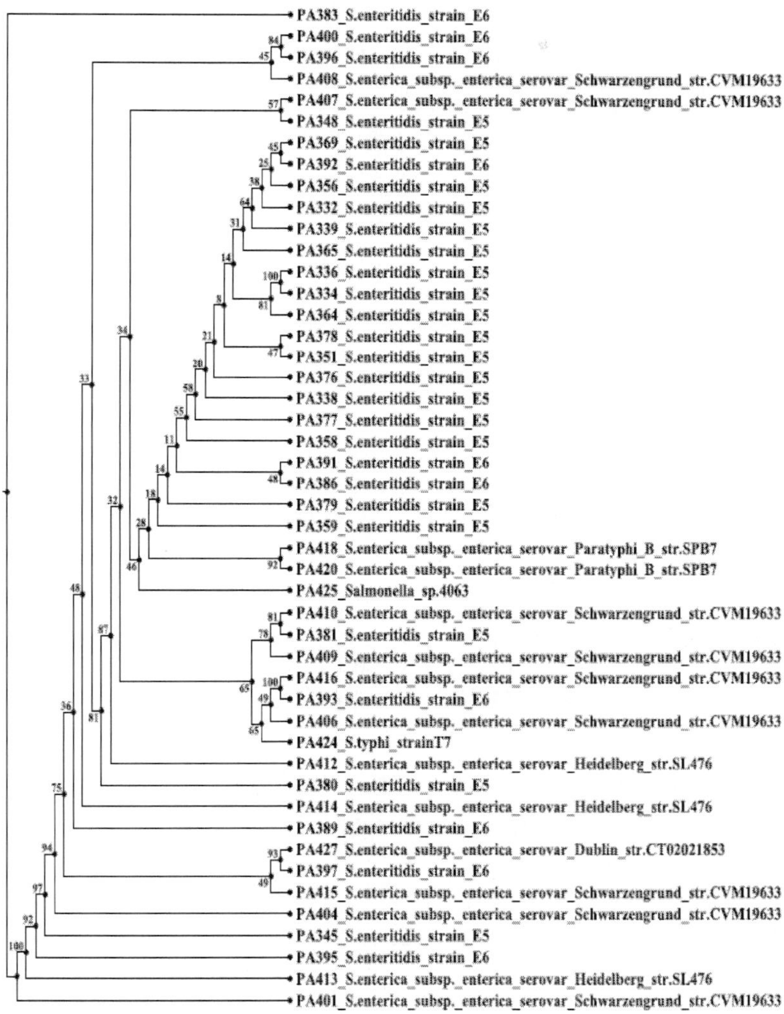

Figure 1: Phylogenetic tree of 16S rRNA gene sequences clustered using the UPGMA method. The numbers at the nodes represent percentages of bootstrap sampling.

Figure 2: Cluster analysis tree of *Salmonella* based on metabolic fingerprinting obtained by Biolog.

Table 1: Biofilm formation by *Salmonella* strains. (a) *Salmonella* strains with high potential to form biofilm; (b) *Salmonella* strains with low potential to form biofilm; (c) *Salmonella* strains unable to form biofilm.

(a) *Salmonella* strains with high potential to form biofilm.

Sample No.	Strain identity
PA348	*S. enteritidis* strain E5
PA339	*S. enteritidis* strain E5
PA351	*S. enteritidis* strain E5
PA338	*S. enteritidis* strain E5
PA377	*S. enteritidis* strain E5
PA345	S. enteritidis strain E5
PA391	*S. enteritidis* strain E6
PA393	*S. enteritidis* strain E6
PA389	*S. enteritidis* strain E6
PA397	*S. enteritidis* strain E6
PA395	*S. enteritidis* strain E6
PA383	*S. enteritidis* strain E6
PA412	*S. enteric serovar Heidelberg*
PA413	*S. enteric serovar Heidelberg*
PA425	*Salmonella sp. 4063*

Table 1: Continued.

(b) *Salmonella* strains with low potential to form biofilm.

Sample No.	Name of bacteria from BLAST
PA369	*S. enteritidis* strain E5
PA336	*S. enteritidis* strain E5
PA334	*S. enteritidis* strain E5
PA364	*S. enteritidis* strain E5
PA358	*S. enteritidis* strain E5
PA379	*S. enteritidis* strain E5
PA400	*S. enteritidis* strain E6
PA396	*S. enteritidis* strain E6
PA392	*S. enteritidis* strain E6
PA415	*S. enteric* serovar Schwarzengrund
PA401	*S. enteric* serovar Schwarzengrund
PA407	*S. enteric serovar Schwarzengrund*
PA410	*S. enteric serovar Schwarzengrund*
PA418	*S. enteric serovar Paratyphi B*
PA420	*S. enteric serovar Paratyphi B*
PA424	*S. typhi strain T7*
PA414	*S. enteric serovar Heidelberg*
PA427	*S. enteric serovar Dublin*

(c) *Salmonella* strains unable to form biofilm.

Sample No.	Name of bacteria from BLAST
PA408	*S. enteric* serovar Schwarzengrund
PA409	*S. enteric* serovar Schwarzengrund
PA406	*S. enteric* serovar Schwarzengrund
PA404	*S. enteric* serovar Schwarzengrund
PA365	*S. enteritidis* strain E5
PA378	*S. enteritidis* strain E5

strains. The biofilm studies (Table 1) showed that the majority of E6 (60% - Table 1 (a)) and Heidelberg (66.7% - Table 1 (b)) strains were able to form biofilm with different potentials, while a very low potential to form biofilm was displayed by 78% of Schwarzengrund strains.

5 Conclusions

Some *Salmonella* strains demonstrated high potential to form biofilm while other strains showed low potential to form biofilm. Biofilm formation potential was not correlated with the metabolic or the genetic background of the tested strains.

Acknowledgements

I would to thank the College of Graduate Studies (CGS) and Research Administration- Kuwait University (grant no. YM06/09) for funding the current project. Many thanks to the Biotechnology Center (BTC) – College of Science, Kuwait University for the provision of the sequencing facility.

References

[1] Cheah, Y. Salleh, N. Lee, L. Radu, S. Sukardi, S. and Simm J. Comparison of PCR fingerprinting techniques for the discrimination of *Salmonella enterica* subsp. *enterica* serovar Weltevreden isolated from indigenous vegetables in Malaysia. *World Journal of Microbiology and Biotechnology*, **24(3)**:327-335, 2008.

[2] D'Aoust, JY. Salmonella and the international food trade. *Int J Food Microbiol.*, **24(1-2)**:11-31, 1994.

[3] Foley, SL. and Lynne, AM. Food animal-associated Salmonella challenged: pathogenicity and antimicrobial resistance. *Journal of Animal Science*, **86 (14)**:E1 73-87, 2008.

[4] Jewes, L. A. Antimicrobial therapy of non-typhi *salmonella* and *shigella* infection. *Journal of Antimicrobial Chemotherapy,* **19**:557-560, 1987.

[5] Jones, BD. Ghori, N. Falkow, S. *Salmonella typhimurium* initiates murine infection by penetrating and destroying the specialized epithelial M cells of the Peyer's patches. *Journal of Experimental Medicine*, **180**: 15–23, 1994.

[6] Jones, BD. Lee, CA. and Falkow, S. Invasion by Salmonella typhimurium is affected by the direction of flagellar rotation. *Infect Immun.*, **60**:2475-2480, 1992.

[7] Khoramian, T. Haryama, S. Kutsukake, K. and Pechere, JC. Effect of motility and chemotaxis on the invasion of Salmonella typhimurium into HeLa cells. *Microbial Pathogene*, **9**:47-53, 1990.

[8] Kingsley, R. Baumler A. and Oelschlaeger & Hacker, J. (eds). Salmonella interactions with professional phagocytes in bacterial invasion into eukaryotic cells. New York: Kluwer Academic/Plenum, pp.321–342, 2000.

[9] Kuske, C. R. Barns, S. M. and Buschm J. D. Diverse uncultivated bacterial groups from soils of the arid southwestern United States that are present in many geographical regions App. Environ. Microbiol. 63:3614-3621, 1997.

[10] Lauren, M. Junker and Cladrym J. High-throughput screens for small-molecule inhibitors of *Pseudomonas aeruginosa* biofilm development. *Antimicrobial Agents and Chemotherapy*, **51(10)**:3582-3590, 2007.

[11] Ohl, M. E. & Miller, S. I. *Salmonella*: a model for bacterial pathogenesis. *Annu Rev Med*, **52**:259–274, 2001.

[12] Rescigno, M. Urbano, M. Valzasina, B. Francolini, M. Rotta, G. et al. Dendritic cells express tight junction proteins and penetrate gut epithelial monolayers to sample bacteria. Nat Immunol., **2**: 361–367, 2001.

[13] Robertson, J. Grant, G. Allen, E. Woodward, M. Pusztal, A. and Flint, H. Adhesion of Salmonella enteerica var Enteritidis strains lacking fimbriae and flagella to rat ileal explants cultured at the air interface or suberged in tissue culture medium. *Journal Medical Microbiology,* **49**:691-696, 2000.

[14] Robertson, J. McKenzie, N. Duncan, M. Vercoe, E. Woodward, M. Flint, A. and Grant, G. Lack of flagella disadvantages Salmonella enterica serovar Enteritidis during the early staged of infection in the rat. *Journal of Medical Microbiology,* **52**:91-99, 2003.

[15] Tsolis, R. Kingsley, R. Townsend, S. Ficht, T. Adams, L. & Baumler A. Of mice, calves, and men. Comparison of the mouse typhoid model with other Salmonella infections. *Adv Exp Med Biol*, **473**:261–274, 1999.

[16] Uyttendaele, MR. Debevere, JM. Lips, RM. Neyts, KD. Prevalence of *Salmonella* in poultry carcasses and their products in Belgium. *Int J Food Microbiol.*, **40**:1–8, 1998.

[17] Vazquez-Torres, A. Jones-Carson, J. Baumler, AJ. Falkow, S. Valdivia, R. et al. Extraintestinal dissemination of *Salmonella* by CD18-expressing phagocytes. *Nature*, **401**: 804–808, 1999.

[18] Vernacchio, V. Vezin, R. Mitchel, A. Lesk, S. Plaut, A. Acheson, D. Characteristics of Persistent Diarrhea in a Community-Based Cohort of Young US Children. *Journal of Pediatric Gastroenterology and Nutrition*, **43(1)**: 52-58, 2006.

[19] United States Food and Drug Administration (FDA) http://www.foodinfonet.com/publication/fdaBAM.htm

Section 2
Ecosystem health

Hazardous substances in the water, biota and sediments of the North Estonian coastal sea

O. Roots[1,2] & Ü. Suursaar[1]
[1]*Estonian Marine Institute, University of Tartu, Estonia*
[2]*Estonian Environmental Research Centre, Estonia*

Abstract

The paper gives an overview of Estonian experiences with hazardous substances monitoring, as well as some recent results on this kind of monitoring in the aquatic environment, biota and sediments of the heavily industrialized North Estonia. Owing to its specific local conditions, the present list of priority hazardous substances of Estonia includes 1- and 2-basic phenols, heavy metals (Cr, Cu, Ni, Zn) and persistent organic pollutants (HCH, HCB, PCB, DDT, PCDD, PCDF, DL-PCB). The concentrations of hazardous substances in the sea and surface waters are currently low, but elevated concentrations may appear in bottom sediments of river estuaries and in fat tissues of fish. During recent decades the state of the North Estonian environment regarding the hazardous substances has continuously improved due to a decreased pollution load, ban of certain compounds, self-purification processes and sinks.
Keywords: priority hazardous substances, POPs, monitoring, bioaccumulation, oil shale, Baltic Sea.

1 Introduction

The Action Plan of the Helsinki Commission (HELCOM) to reduce pollution of the Baltic Sea and restore its ecological status by 2020 was approved in November 2007 by representatives of all the nine countries surrounding the Baltic Sea. One of the four main topics, the hazardous substances segment, defines the main goals as follows: to reach the concentrations of hazardous substances close to background levels for naturally occurring substances, close to zero for manmade substances, have all fish become safe to eat and reach pre-Chernobyl levels of radioactivity in the Baltic Sea ecosystem [1].

WIT Transactions on Ecology and the Environment, Vol 132, © 2010 WIT Press
www.witpress.com, ISSN 1743-3541 (on-line)
doi:10.2495/ETOX100081

In the heavily industrialized northern part of Estonia, oil shale based chemical industry and power plants, cement and pulp mills, and a Soviet-era depository of radioactive wastes have seriously impacted virtually every component of the environment. It is also the major pollution source and a "hot spot" for the whole Baltic Sea environment [2,3]. Therefore, the main efforts in the monitoring of hazardous substances in Estonia have been concentrated into that area.

It is known that the list of modern time chemicals and ingredients, which could be viewed as pollutants, is enormously long. Some of the pollutants may be quite common in the environment (e.g. so-called nutrients, such as N and P salts, oils, heat) and become dangerous only due to excessive anthropogenic emissions. However, hardly degradable synthetic compounds can be hazardous even in very low concentrations. The main criteria for deciding their hazard are toxicity (both acute and chronic), persistency, bio-accumulating properties, and some other aspects (e.g. carcinogenic or synergic influences) [1,4]. As far as it is practically possible to monitor only selected substances, one should also estimate the amount of their emissions and spread.

The paper focuses mainly on the so-called toxic priority substances, which were initially listed in the Stockholm Convention and in UNEP 2003 protocols [4,5]. However, the list itself is constantly changing. Also, as far as both widespread and more local ingredients exist, each country has to identify their own versions of the list. In 1999–2001, three nation-wide inventories of hazardous substances were performed in Estonia. The protocols of UNEP require a ban or minimized use of these priority hazardous substances. Import of chlororganic pesticides to Estonia has been prohibited since 1967 and Estonia itself has not been manufacturing them. The rest of the obsolete pesticides have been disposed of in Estonia. However, both persistent organic pollutants (POPs) and heavy metals (HMs) still circulate in the environment and food webs. Discharges of certain compounds inevitably continue, although in reduced quantities, and new compounds are being synthesised. Therefore, the horizon of knowledge on that issue is also broadening and modifying.

The aims of the paper are to (1) give a brief overview of Estonian hazardous substances monitoring activities and corresponding environmental management, and (2) to summarise the results of the monitoring of hazardous substances in the water, biota and sediments of the North Estonian coastal environment.

2 Material and methods

2.1 Historic overview of monitoring

For analysis of water quality in the North Estonian coastal sea we used the monitoring databases of the Estonian Marine Institute and Estonian Environmental Research Centre. They include both the historical data over the period 1968–1993 [6,7], as well as the more recent data from the new revised national monitoring programme [8,9].

Regular pollution control in some parts of the coastal waters of Estonia began in 1967, after the endorsement of the General Plan for Water Use and Protection of the USSR. Until 1991, the marine monitoring was carried out in about 40–50 stations by the former Estonian Hydrometeorological Service within the monitoring system of the Soviet Union. The variables measured included hydrological variables and nutrients, as well as oils, phenols, detergents, POPs, lignine, and some trace metals.

Starting from 1991, revision of the national monitoring began. Evaluation of the national monitoring data revealed excessive variability, systematic biases and incompatibility with other data sets due to different sampling and analysing standards [6]. The present programme is based on the requirements of the Baltic Monitoring Programme from 1998 (COMBINE), and the European Water Framework Directive. It includes several sub-programmes, one being the monitoring of priority hazardous substances in the marine environment, which is currently conducted by the Estonian Marine Institute.

2.2 Material and methods for hazardous substances monitoring

In Estonia, the concentrations of hazardous substances in the aquatic environment have been surveyed since 1974 [7,8]. Since then, both the list of substances monitored and the locations and environs have changed several times. As the concentrations of POPs in the coastal waters appeared mostly to be below detection level of the analytical methods of these days, the focus of the survey shifted to the selected fish species in the coastal sea and rivers. As far as many chemical contaminants become concentrated at the top of the food chain, such monitoring offers more stable values and an integrated view together with food safety issues.

Estonia specified its list of hazardous substances for surface water on national level proceeding from the directive 92/446/EEC of the European Council from 1992 and Commission decision 95/337/EC from 1995. At present, the list of priority hazardous substances of Estonia includes 1-basic phenols (p-, m- and o-cresols; 2,3-, 2,6-, 3,4- and 3,5-dimethylphenols), 2-basic phenols (resorcinol, 5-methyl resorcinol and 2,5-dimethyl resorcinol), Ba (only in Vend-layer groundwater), heavy metals (Cr, Cu, Ni, Zn) together with their compounds and persistent organic pollutants (HCH, HCB, PCB, DDT, PCDD, PCDF, DL-PCB) [10]. Chlororganic substances in certain fish and mollusc (*Macoma, Saduria*) species are sampled annually in three monitoring clusters (in the bays of Pärnu, Tallinn and Kunda). According to HELCOM recommendations, the selected bio-indicator is the female Baltic herring of two-three years age [11]. In selected North Estonian rivers, samples of fishes (or sediments, in case of intakes) are performed. The method of so-called rotation monitoring is used. I.e., not all of the substances and locations are monitored each year.

Most of the water and sediment samples have been taken by employees of the Estonian Environmental Research Centre (EERC), which also made the most laboratory tests. The Estonian Accreditation Centre certifies that EERC has competence according to ISO/IEC 17025:2005 to conduct tests in the field of water, soil, food, air and petroleum product analyses [8]. More detailed description of sampling techniques as well as analytical procedures can be found in [12–14]. The biota samples have been taken by employees of Estonian Marine

Institute and some water samples from North-Estonian rivers and estuaries by Tallinn Technological University [15]. The analyses for polychlorinated dibenzo-p-dioxins (PCDDs), polychlorinated dibenzofurans (PCDFs) and dioxin-like polychlorinated biphenyls (DL-PCBs) were ordered from Finland.

3 Study area

The North Estonian coastal sea stretches for about 300 km from west to east along the Gulf of Finland (Fig. 1), which has an area of 29 500 km^2 and average depth of 37 m. The Gulf receives a relatively large pollution load both from rivers (Neva, Narva, Kymijoki), as well as from industries and municipalities [2,3]. The circulation scheme in the Gulf of Finland is mostly wind-driven and although certain statistical long-term patterns can be found [16], it is quite variable in time and space. The northern coast of Estonia has a relatively straight coastline where good mixing conditions due to waves and currents prevail [17].

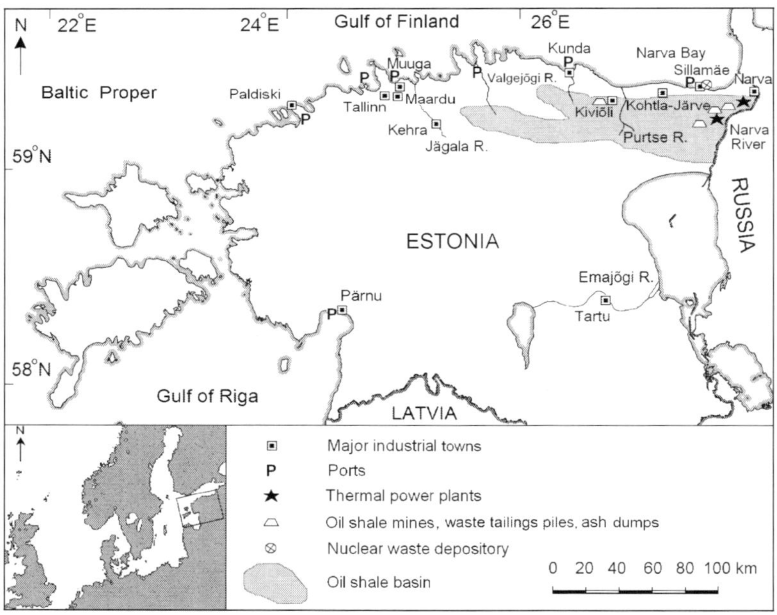

Figure 1: The study area: North Estonian industrial zone and the coastal sea.

As natural degradation of hazardous substances is low, the main losses occur via sinks to bottom sediments, fish catches, as well as by simple dilution and spread over larger areas, (if the concentrations are lower there). On the other hand, the persistent pollutants can also be brought to the site by rivers, sea-currents and atmospheric deposition.

Bearing in mind the fact that the most of the Baltic Sea organisms live at the edge of their physiological tolerance range, anthropogenic chemical pollution has to be seen as a further stress factor acting upon marine biodiversity. Although some blue-green algae produce natural toxins in substantial quantities, which can affect, for instance, reproduction and growth of fish [18], the anthropogenic (synthetic) toxic and hazardous substances vastly dominate in the marine environment. Despite the recent reduction in pollution load and good hydrodynamic dilution conditions, the area has not yet fully recovered from the past anthropogenic pressure [2] and some persistent compounds are bioaccumulated in the biota and bottom sediments [4,11].

4 The main pollution sources

Oil shale mining and oil shale-based power and chemical industry has been the main cornerstone of the Estonian economy for decades. On the other hand, it is also the largest source of pollution. Estonian oil shale (kukersite) reserves are vast (Table 1). The majority of the mined oil shale (10–15 million tons/yr) is used for electricity generation in two thermal power plants near Narva and Kohtla-Järve with total capacity of 2.4 MW (Fig. 1). As oil shale is a low-grade fossil fuel, each year about 4–5 million tons of oil shale ash and semi-coke is dumped near the power plants, where residual organic matter is prone to self-ignition and give gaseous emissions (e.g. SO_2, NOx), and influencing also rivers of Narva, Purtse, Valgejõgi and Jägala (Fig. 1).

For the coastal sea, however, the main risks appear from the effluents of oil shale-based chemical industry and oil extraction. During the economic peak of 1980s, Kohtla-Järve and Kiviõli chemical companies discharged substantial amounts of phenols (650 t/yr), oils, and different hazardous substances to the rivers and coastal sea [2,3,15].

Table 1: Resources and extraction of Estonian oil shale in million tons [19].

Extraction		Reserves and resources (2007)		
2005	2006	Proved mineable reserves	Inferred mineable resources	Sub-marginal mineral resources
11.3	12.0	1129.2	268.6	3502.7

Due to a more than two-fold decrease in production, as well as construction of wastewater treatment plants over the last two decades, the pollution load to the coastal sea has decreased nearly two-fold. Still, in Estonia the generation of hazardous waste per capita in a year (6.4 tons in 2008) is the highest among European countries [20], which is mainly due to the oil-shale industry.

Regarding the other pollution sources in North Estonia, two large pulp and paper mills have discharged, either directly or via rivers, to the Gulf of Finland. The Tallinn mill used to have an annual production of about 68 000 t of sulphite cellulose, and the Kehra mill produced annually 52 000 t of sulphate cellulose. The closing of these mills in the 1990s was largely a result of altered environmental requirements. Today, most pulp mills around the Baltic Sea utilize a chlorine-free bleaching process. However, about 50% of the total organochlorine inputs from pulp mills since the early 1940s still reside in the Baltic Sea – mainly in the bottom sediments [11,21].

On the shore of the Gulf of Finland, a highly specialized chemical and metallurgy plant of Sillamäe (Fig. 1) was established in 1946. For production of nuclear materials, uranium was at first locally mined from dictyonema argillite. In 1970 the plant was switched to processing imported loparite. Uranium enrichment was finished in 1989, since then enrichment of rare metals (such as niobium) continues at AS Silmet. However, since 1959, on the immediate shore of the Gulf of Finland, a 50 ha nuclear waste depository grew. Probably some 1200 tons of uranium, 800 tons of thorium enrichment residuals and other hazardous substances are buried there. The closure and sanitation of the depository in 1998–2008 became one of the highest priority environmental projects over the whole Baltic Sea basin. The leakage through the dam to the sea should be negligible now [22].

5 Results and discussion

5.1 Phenols

While the general European hazardous substances list includes nonylphenoles and octylphenoles [23], the Estonian list specifically includes 1-basic and 2-basic phenols. The phenols are discharged to the coastal sea of Estonia mainly from oil-shale based chemical industry. Studies of phenol concentrations showed that concentrations of 1-basic phenols (p-, m- and o-cresols; 2,3-, 2,6-, 3,4- and 3,5-dimethyl phenols) as well as 2-basic phenols (resorcinol, 5- methyl resorcinol, 2,5-dimethyl resorcinol) are currently low, fluctuating within the range <0.5 – 5 µg/l in North Estonian river estuaries [15]. However, excessively high concentrations of "phenols" (unspecified) were frequently found in the coastal waters of North Estonia in 1970s and 1980s [6].

A regulation of the Ministry of Social Affairs ("Quality and control requirements to surface and ground water used or intended for use for the production of drinking water" from January 2003) specifies upper limits for different quality classes: class 1 – 0.001 mg/l, class II – 0.005 mg/l and class III – 0.1 mg/l. According to these standards, all the studied lower reaches and estuaries of the rivers have belonged to I and II classes in recent years. 0.005 mg/l is also the upper limit for the European recreational (bathing) water, which is met by most of the Estonian rivers and beaches during summer months [15].

5.2 Heavy metals in water, sediments and biota

Concentrations of trace metals (including so-called heavy metals: Cu, Zn, Hg, Pb, Cd) in the seawater have been episodically determined in the Baltic Sea [11], including the Estonian coastal waters [7]. More recently, the following metals are included in EIA, the "extended" list for metals: arsenic (As), cadmium (Cd), chromium (Cr), copper (Cu), mercury (Hg), nickel (Ni), lead (Pb) and zinc (Zn).

Table 2: Concentrations of dissolved metals (ng/kg) in the North Atlantic [24,25] and the Baltic Sea [26] along with the factor, showing the elevated Baltic Sea concentrations.

Metal	North Atlantic	Baltic Sea	Factor
Mercury (Hg)	0,15–0,3	5–6	~ 20–40
Cadmium (Cd)	4 (+/-2)	12–16	~ 3–4
Lead (Pb)	7 (+/-2)	12–20	~ 2–3
Copper (Cu)	75 (+/-10)	500–700	~ 7–10
Zinc (Zn)	10–75	600–1000	~ 10–40

However, in the framework of COMBINE, the status for seawater analyses is still "tentative". Such analyses usually require high-cost instrumentation; the concentrations are extremely low, as is the relative preciseness, while the data scatter is remarkable. In 1980s and 1990s, the typical concentration for dissolved Cd was 0.1–0.3 nmol/dm^3, 5–15 nmol/dm^3 for Cu and 10–30 nmol/dm^3 for Zn [11], the particulate concentrations being 5–10 fold lower.

Despite some decline in loading, the concentrations in the Baltic Sea are still 3 to 50 times higher than in the North Atlantic (Table 2). Heavy metals can reach the marine environment via the atmosphere, through discharges or natural drainage. As a direct impact, annual emissions from the Baltic Sea countries decreased in 1996–2000, by 26% for cadmium, 15% for mercury and lead [22]. Concentrations of Cd, Pb and Zn are on average higher in the south-western parts of the Baltic Sea, where atmospheric deposition is greater. One fifth of the cadmium input to the Baltic Sea comes from atmospheric deposition, carried by the prevailing south-westerly winds. Atmospheric emissions of Estonian power plants have declined from 1 t to 0.7 t in 1997–2000 [8].

Determination of the concentrations in the fish tissues and bottom sediments is more reliable and informative. According to the Estonian hazardous substances monitoring results in 2001–2007, the typical concentrations of Hg in the herring tissues were 0.02–0.04 mg/kg (dry weight), 0.4–1.5 mg/kg for Cd, 0.2–0.3 mg/kg for Pb, 60–100 mg/kg for Zn and 4–16 mg/kg for Cu. All the respective concentrations are generally lower than in 1990s, but Cu and Pb still needs some special attention [9].

In addition to the top of the food-web, persistent contaminants tend to concentrate into sediments of the deeper marine areas (Fig. 2). In that sense, the semi-enclosed Baltic Sea acts as a separate, isolated from the rest of the World Ocean basin with large inputs and only internal sinks. Regarding the metals in the sediments of the Gulf of Finland, also vicinity to the historical pollution source plays an important role. The quality of sediments in the Gulf of Finland has been classified using the criteria by SEQC, the Swedish Environmental Protection Agency (Table 3).

According to this classification, the state of surface sediments in the Gulf of Finland was not satisfactory, especially in the NE and E section of the Gulf of Finland. The sediments were "significantly" or "largely" polluted with heavy

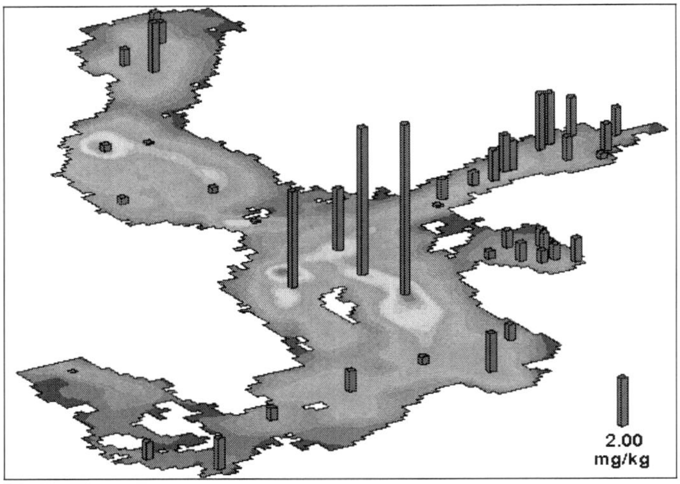

Figure 2: Spatial distribution of heavy metal cadmium in bottom sediments of the Baltic Sea, 0–2 cm layer (modified from [11]).

Table 3: Sediment quality classification (SEQC, mg/kg dry weight [27]).

Metal	Class 1 Little/none	Class 2 Slight	Class 3 Significant	Class 4 Large	Class 5 Very large
Pb	<31	31–47	47–68	68–102	>102
Hg	<0.04	0.04–0.10	0.10–0.27	0.27–0.72	>0.72
Cu	<15	15–30	30–60	60–120	>120
Cd	<0.2	0.2–0.5	0.5–1.2	1.2–3	>3.0
Zn	<85	85–125	125–196	196–298	>298
Cr	<80	80–112	112–160	160–224	>224

metals. The quality of the sediments surprisingly showed only "slight" or "significant" pollution along North Estonian coast [27]. This difference probably appears as a result of shallower sea with good dilution conditions on one hand, and different industry profiles in Estonia, Russia (St. Petersburg) and Finland (Kotka). According to Perttilä [28], one should bear in mind that over the years, some of these metal deposits will be transformed into hazardous compounds (incl. organic tin and mercury), so their release back into the marine ecosystem may result in very harmful effects.

5.3 Persistent organic pollutants in river deltas and fish

The study by Loos and co-authors [29] provides the first EU-wide reconnaissance of the occurrence of polar organic persistent pollutants in European river waters. The selection of sampling sites has done by the participating EU Member States. 122 individual water samples from 27 European countries were analysed for 35 selected compounds (HCH, HCB, PCB,

DDT, PCDD, PCDF, DL-PCB and some more). In that study, only the dissolved (liquid) water phase, and not the suspend material was investigated. The results suggested that just about 10% of the river water samples could be classified as "very clean" in terms of chemical pollution, since they contained only a few compounds in very low concentrations. The most pristine water samples came from Sweden, Lithuania and Estonia (rivers of Narva, Purtse and Emajõgi) [29].

As persistent organic pollutants tend to accumulate into fats, the primary attention has to be paid to high-fat food, especially fish. The results indicate that the concentrations depends both on location of the catch, but also on fish age, sex, maturity of gonad, fat content and some other characteristics. Also temporal variability can be large [13,14], but generally decreasing trends are visible in some POPs (Figs. 3,4). The dioxin (PCDD) content in fish has been quite well studied in Estonia since 2002. In general, no risk is incurred by consuming perch, pike-perch and flounder. Atlantic salmon, sea trout and eel have not been adequately examined, as their proportion in human consumption is relatively small in Estonia. However, consumption of the "large Baltic herring", older than 5 years and with the length of more than 17 cm, should be avoided or constrained, especially by pregnant women [13].

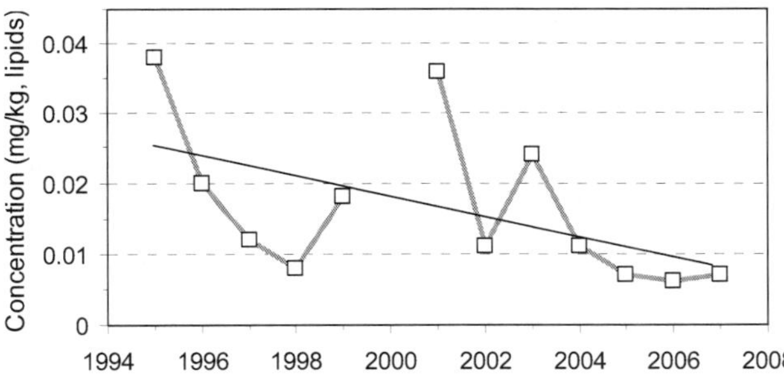

Figure 3: Concentrations of g-HCHs (mg/kg lipids) in the muscle tissues of Baltic herring in the Estonian coastal waters in 1995–2007.

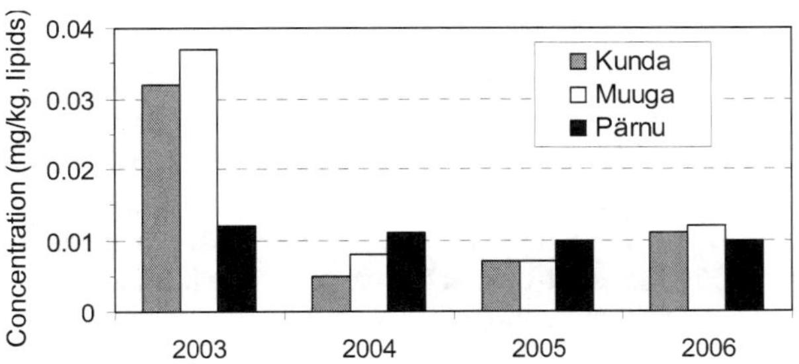

Figure 4: Concentrations of HCBs in the muscle tissues of Baltic herring in different sampling clusters of Estonian coastal sea in 2003–06 [9,14].

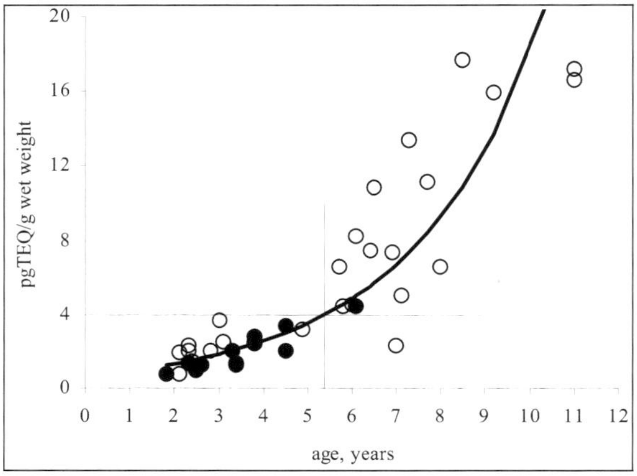

Figure 5: Age-dependent dioxin concentrations in the Baltic herring muscle tissues of Estonian (black circles) and Finnish (white circles) catches [13].

Dioxin concentration grows exponentially with the age (Fig. 5). Currently the age of the Baltic herring in the coastal waters of Estonia is mostly 2–4 years and the proportion of the fish catches older than 5 years is 10% [13,14]. Dioxins could also be found in the bottom sediments. Thus, extensive disturbance of the seabed in the Gulf of Finland by dredging or installing pipelines could result in a massive release of dioxins and other contaminants into the foodweb [28].

6 Conclusions

Monitoring programmes of hazardous substance may vary from one country to another, as well as in time. At present, the list of priority hazardous substances of Estonia includes 1-basic and 2-basic phenols, Ba (only in groundwater), heavy metals (Cr, Cu, Ni, Zn) and persistent organic pollutants (HCH, HCB, PCB, DDT, PCDD, PCDF, DL-PCB).

The concentrations of hazardous substances in most Estonian surface and coastal waters are currently low. However, as these substances are hardly degradable, they tend to accumulate in bottom sediments and bio-accumulate in food-webs, especially in fat tissues of fish and mammals. Baltic Sea acts as an isolated basin, where large pollution load from surrounding countries could be counteracted mainly by internal sinks. The effect of bans of certain substances, as well as other pollution control mechanisms, is not immediately visible. For instance, due to elevated dioxin and heavy metal concentrations, some fish (large herring, cod) are not advised to eat frequently or in large quantities.

Long-term improvement of the environment quality in North Estonia due to decreased pollution load and self-purification is still evident in recent decades. It also reflects the endorsement of rigid pollution control mechanisms, efficient monitoring and improvement in environmental legislation.

Acknowledgements

The study was supported by the ESF grant 7609 and the theme SF0180104s08.

References

[1] HELCOM. *Baltic Sea Action Plan.* Adopted on 15 November 2007 in Krakow, Poland by the HELCOM Extraordinary Ministerial Meeting. http://www.helcom.fi/stc/files/BSAP/BSAP_Final.pdf. 2007.

[2] Suursaar, Ü., Aps, R., Kotta, I. & Roots, O., North-East Estonian coastal sea: recovery from the past anthropogenic pressure and new stressors on the background of natural variability. *Ecosystems and Sustainable Development VII. Book Series: WIT Transactions on Ecology and the Environment*, **122**, WIT Press: Southampton, Boston, pp. 331–342, 2009.

[3] HELCOM. The Fourth Baltic Sea Pollution Load Compilation (PLC-4). *Balt. Sea Environ. Proc.*, **93,** 198 pp., 2004.

[4] HELCOM. Hazardous substances of specific concern to the Baltic Sea. Final report of the HAZARDOUS project. *Balt. Sea Environ. Proc.*, **119**, 96 pp., 2009.

[5] UNEP. *Europe Regional Report 2002. Regionally Based Assessment of Persistent Toxic Substances*. UNEP Chemicals: Geneva, 2003.

[6] Suursaar, Ü., Estonian marine monitoring 1968–1991: Results and evaluation. *Finnish Marine Research*, **262**, pp. 123–134, 1994.

[7] Jankovski, H., Simm, M. & Roots, O., Harmful substances in the ecosystem of the Gulf of Finland. Part I, Trace metals. *EMI Report Series,* **4**, Tallinn, 158 pp., 1996.

[8] Roose, A. & Roots, O., Monitoring of priority hazardous substances in Estonian water bodies and in the coastal Baltic Sea. *Boreal Env. Res.*, **19**, pp. 89–102, 2005.

[9] Simm, M., Hazardous substances in the coastal sea. *Estonian Environmental Monitoring 2004–2006,* ed. K. Väljataga, Ministry of Environment: Tallinn, pp. 77–79, 2008.

[10] Roots, O., Proposal for selection of national priority hazardous substances for Estonian surface water bodies. *Ecological Chemistry. St. Petersburg University and Thesa*, **17** (1), pp. 22–34, 2008.

[11] HELCOM. Fourth Periodic Assessment of the State of the Marine Environment of the Baltic Sea Area, 1994–1998. *Balt. Sea Environ. Proc.*, **82B**, 218 pp., 2002.

[12] Roots, O., Halogenated environmental contaminants in fish from Estonian coastal areas. *Chemosphere*, **43** (4–7), pp. 623–632, 2001.

[13] Roots, O., Lahne, R., Simm M. & Schramm, K.W., Dioxins in the Baltic herring and sprat in Estonian coastal waters. *Organohalogen Compounds*, **62**, pp. 201–203, 2003.

[14] Lukki, T., Roots, O., Simm, M., Talvari, A. & Tuvikene, A., PCBs, HCHs and HCB in the Baltic Sea herring of the Estonian coastal sea. *Organohalogen compounds*, **70**, pp. 2098–2101, 2008.

[15] Hannus, M., Leisk, Ü. & Loigu, E., Hazardous substances in rivers. *Hazardous substances in Estonian Environment,* eds. A. Roose, E. Otsa, O. Roots, Tartu University Publishing House: Tartu, pp. 35–37, 2003.

[16] Andrejev O., Myrberg K., Alenius P. & Lundberg P., Mean circulation and water exchange in the Gulf of Finland – a study based on three-dimensional modelling. *Boreal Env. Res.,* **9**, pp. 1–16, 2004.

[17] Suursaar, Ü., Aps, R., Martin, G., Põllumäe, A. & Kaljurand, K., Monitoring of the pulp mill effluents in the coastal waters of North Estonia. *Water Pollution I.* Book Series: *WIT Transactions on Ecology and the Environment*, **111**, WIT Press: Southampton, Boston, pp. 217–226, 2008.

[18] Ojaveer, E., Simm, M., Balode, M., Purina, I. & Suursaar, Ü., Experiments on the effect of *Microcystis aeruginosa* and *Nodularia spumigena* upon the survival of Eurytremora affinis and fertilization, embryonic and larval development of Baltic herring *Clupea harengus* membras. *Environmental Toxicology*, **18** (4), pp. 236–242, 2003.

[19] Raudsep, R., Estonian geo-resources in the European context. *Estonian Journal of Earth Sciences*, **57,** pp. 80–86, 2008.

[20] Rohtla, R. (ed.), *Monthly bulletin of Estonian statistics, 12/08.* Statistics Estonia: Tallinn, 148 pp., 2009.

[21] Kankaanpää, H., Lauren, M., Saares, R., Heitto, L. & Suursaar, Ü., Distribution of halogenated organic material in sediments from anthropogenic and natural sources in the Gulf of Finland catchment area. *Environmental Science and Technology*, **31** (1), pp. 96–104, 1997.

[22] HELCOM. Radioactivity in the Baltic Sea 1992–1998. *Balt. Sea Environ. Proc.*, **85**, 102 pp., 2003.

[23] HELCOM. Heavy Metal Pollution to the Baltic Sea in 2004. *Balt. Sea Environ. Proc.*, **108**, 33 pp., 2007.

[24] Dalziel, J., Reactive mercury in the eastern North Atlantic and Southeast Atlantic, *Marine Chem.*, **49**, pp. 307–314, 1995.

[25] Kremling, K. & Streu, P., The behaviour of dissolved Cd, Co, Zn, and Pb in North Atlantic near-surface waters (30°N/60°W to 60°N/2°W). *Deep Sea Research*, **48**, pp. 2541–2567, 2001.

[26] Pohl, C. & Hennings, U., The coupling of long-term trace metal trends to seasonal diffusive trace metal fluxes at the oxic-anoxic interface in the Gotland Basin; (57°19,20`N; 20°03,00´E) Baltic Sea. *Journal of Marine Systems*, **56**, pp. 207–225, 2005.

[27] Assessment of Environmental Quality - Coasts and Sea. *Naturvårdsverket Rapport,* No. 4914, 134 pp., 1999.

[28] Perttilä, M., Kankaanpää, H., Kotilainen, A., Laine, A., Lehtoranta, J., Leivuori, M., Myrberg, K. & Stipa, T., Implementation of the North European gas pipeline project – data inventory and further need for data for environmental impact assessment. *MERI*, **58**, 22 pp., 2006.

[29] Loos, R., Gawlik, B.M., Locoro, G., Rimaviciute, E., Contini, S. & Bidoglio, G., *EU Wide Monitoring Survey of Polar Persistent Pollutants in European River Waters*. European Commission, EUR 23568EN, 2008.

Controlling groundwater pollution from petroleum products leaks

M. S. Al-Suwaiyan
Civil Engineering Department,
King Fahd University of Petroleum and Minerals, Saudi Arabia

Abstract

Groundwater is the main source of potable water in many communities. This source is susceptible to pollution by toxic organic compounds resulting from the accidental release of petroleum products. A petroleum product like gasoline is a mixture of many organic compounds that are toxic at different degrees to humans. These various compounds have different characteristics that influence the spread and distribution of plumes of the various dissolved toxins. A compositional model utilizing properties of organics and soil was developed and used to study the concentration of benzene, toluene and xylene (BTX) in leachate from a hypothetical site contaminated by BTX. Modeling indicated the high and variable concentration of contaminants in leachate and its action as a continuous source of groundwater pollution. In a recent study, the status of underground fuel storage tanks in eastern Saudi Arabia and the potential for petroleum leaks was evaluated indicating the high potential for aquifer pollution. As a result of such discussion, it is concluded that more effort should be directed to promote leak prevention through developing proper design regulations and installation guidelines for new and existing service stations.

Keywords: groundwater pollution, petroleum products, dissolved contaminants, modelling contaminant transport.

1 Introduction

Water covers about 73% of our planet with a huge volume of 1.4 billion cubic kilometers most of which is saline. According to the water encyclopedia [1], only about 3-4% of the total water is fresh. Most of the freshwater exist as ice in the polar region leaving about 9 million cubic kilometers of fresh water existing

WIT Transactions on Ecology and the Environment, Vol 132, © 2010 WIT Press
www.witpress.com, ISSN 1743-3541 (on-line)
doi:10.2495/ETOX100091

as groundwater and surface water. Looking at numbers one will realize that most fresh water on earth exists as groundwater making it arguably the most valuable resource on earth.

The value of this resource can be reduced drastically if its quality deteriorates. A common source for groundwater pollution is hydrocarbon leaks from underground storage tanks which happen due to improper design, human error, accidents, or simply due to the natural aging and deterioration of the tank itself or its associated piping and fittings.

An underground storage tank (UST) or its associated piping may leak, releasing a certain volume of hydrocarbon into the subsurface. Depending on the spill volume, type and subsurface properties, the hydrocarbon may be trapped in the unsaturated zone. For high spill volume, it will continue to migrate down reaching close to the water table. The mobile phase near the water table can migrate laterally in the same direction as groundwater. Part of the hydrocarbon will dissolve slowly in the groundwater providing a long term source of groundwater contamination by means of a contaminated plume that grows in size with time [2].

The major steps involved in dealing with spills and trying to restore the subsurface have to do, in the initial phase, with source control and development of thorough understanding of the subsurface conditions and the extent of contamination which should be followed by intensive use of modeling techniques in order to examine and select the most effective means of aquifer restoration and to examine the system behavior under various possible scenarios.

2 Contamination assessment and monitoring

Field investigation at this stage aims at assessing the extent of contamination and knowing the distribution of the released contaminants. It may involve sampling of aquifer material, construction of wells screened in hydrocarbon zone and wells screened below the water table, which can provide information such as thickness of hydrocarbon in wells, concentration of dissolved contaminants as well as approximate water table elevation. These are the primary data that must be used to evaluate nature and extent of groundwater pollution. Soil samples collected during the field investigation can be taken to the laboratory to get their grain size distribution which may be in turn used in models such as the one presented by Mishra et al. [3] to generate a first approximation for the hydraulic properties of the subsurface. A review for estimating spill volume is presented by Saleem et al. [4].

It is well established that monitoring wells are not reliable for spill detection and quantification since in many field cases, leaks are accidentally discovered by detecting free product in utility manholes not by finding free product monitoring wells.

3 Distribution of contaminants

Farr et al. [5] as well as Lenhard and Parker [6] showed that the vertical distribution of a hydrocarbon after a spill is expected to be influenced by the spill volume, soil properties like displacement head, distribution index and value of residual saturation. These properties reflect the grain-size distribution in the aquifer material which varies significantly from one location to another. Hydrocarbon properties also influence its distribution including density, surface tension, viscosity, solubility and volatility. In general a hydrocarbon can exist in either of four classes. It can be held by capillary and adsorptive forces in the unsaturated/saturated zone as residual or immobile phase which approximately remain in its place but slowly dissolving part of its mass with any water flow. It can also exist as a vapor phase in the unsaturated zone or as free phase near the water table and in monitoring wells. Finally it can exist as a dissolved phase in groundwater at relatively very low concentrations but note that most of these products are very harmful to humans even at trace levels.

4 Subsurface near petroleum processing facilities

The subsurface near petroleum processing facilities or aging service station most likely will contain many organic compounds originally existing in oil. Among these compounds are benzene, toluene and xylene (BTX). BTX are common organic pollutants associated with accidental leaks of common fuels. Their presence in the subsurface will result eventually in groundwater contamination which can create a hazard affecting human health. The health effects associated with long term expose to these products according to EPA [7] include chromosome aberrations, cancer, liver and kidney damage, damage to central nervous system, nervous disorders including spasms, tremors, impairment of speck, hearing, vision, memory, coordination. National Primary Drinking Water Regulations [7], give maximum contaminant levels (MCL) of 5 ppb, 1 ppm, 10 ppm, for benzene, toluene and xylene respectively. To assess the degree of possible groundwater pollution from an accidental spill, the characteristics of the leachate from the contaminated zone, mainly pollutant concentrations has to be studied which can be done by developing a compositional model.

5 Modeling leachate characteristics

Consider the hypothetical subsurface near a petroleum processing facility shown in fig. 1. The shaded area near ground surface represents an idealized zone polluted by residual amounts of BTX. As water infiltrates due to precipitation, irrigation or process activities, it will leach these contaminants slowly forming a continuous source of contamination into the underlying unconfined aquifer. In this section a model to study this process will be developed.

5.1 Leaching model development

After a hydrocarbon spill in the vadose zone its components will partition between four possible phases, namely water, air, soil and free hydrocarbon. For each component the concentration in water is related to the bulk concentration through [8]:

$$m_i = B_{w_i} C_{w_i} \tag{1}$$

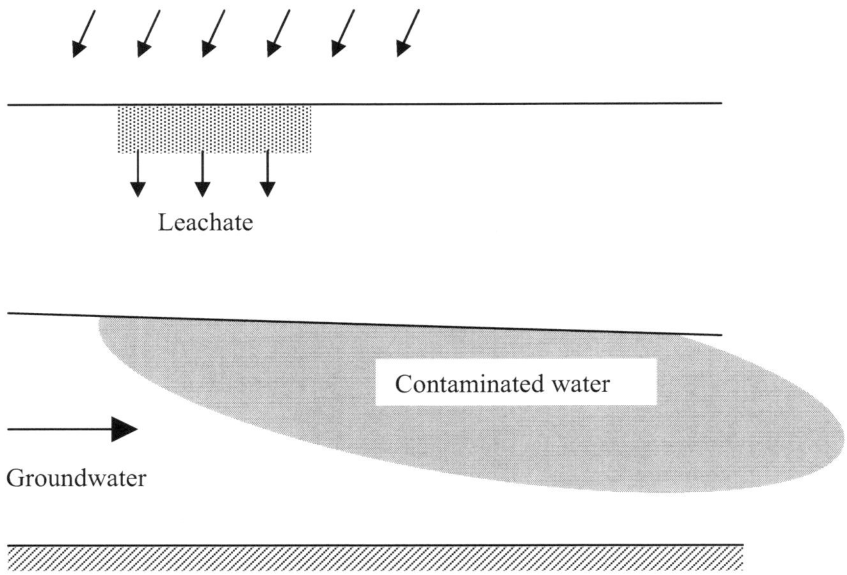

Figure 1: Schematic view for contaminated soil and groundwater aquifer.

where:

m_i = bulk concentration of compound i

B_{w_i} = bulk water partitioning coefficient for compound i

C_{w_i} = concentration of compound i in water

The bulk water partitioning coefficient will be influenced by component volatility, adsorption properties as well as the distribution of compound in the free hydrocarbon.

Applying principle of mass balance will allow us to develop the concentration of pollutants in leachate as a function of time as outlined by Charbenuea [8].

Referring to fig.1, the total mass present for compound i is given by:

$$M_i = A\ L_o\ B_{w_i}\ C_{w_i} \tag{2}$$

where:

M_i = total mass of compound i

A = area of contaminated zone

L_0 = depth of contaminated zone

Neglecting volatility and degradation and assuming mass is lost only with leaching water, the mass balance equation becomes:

$$\frac{d\,M_i}{dt} = \frac{d}{dt}[A\ L_o\ B_{w_i}\ C_{w_i}] = -\,qA\ C_{w_i} \tag{3}$$

The above equations can be solved numerically to come up with the BTX concentrations at various times.

5.2 Modeling results and discussion

The above model was applied to the hypothetical case shown in fig. 1, using typical values for the parameters involved. These values include a porosity of 0.45, a water flow rate of 1 cm/day, and an average moisture content of 0.225, and a residual hydrocarbon content of 0.0675. A contaminated depth of 10 cm was selected and the contaminating mixture consisted of equal volumes of BTX. The model was run to generate the concentration of the three organic compounds in the leachate and their temporal variation.

The characteristics of the leachate can be examined through fig. 2, which indicates that at early times, benzene acts as the dominant pollutant affecting groundwater. Benzene concentration decreases at a relatively sharp rate. After about seven months, toluene takes over as the main pollutant and assumes that

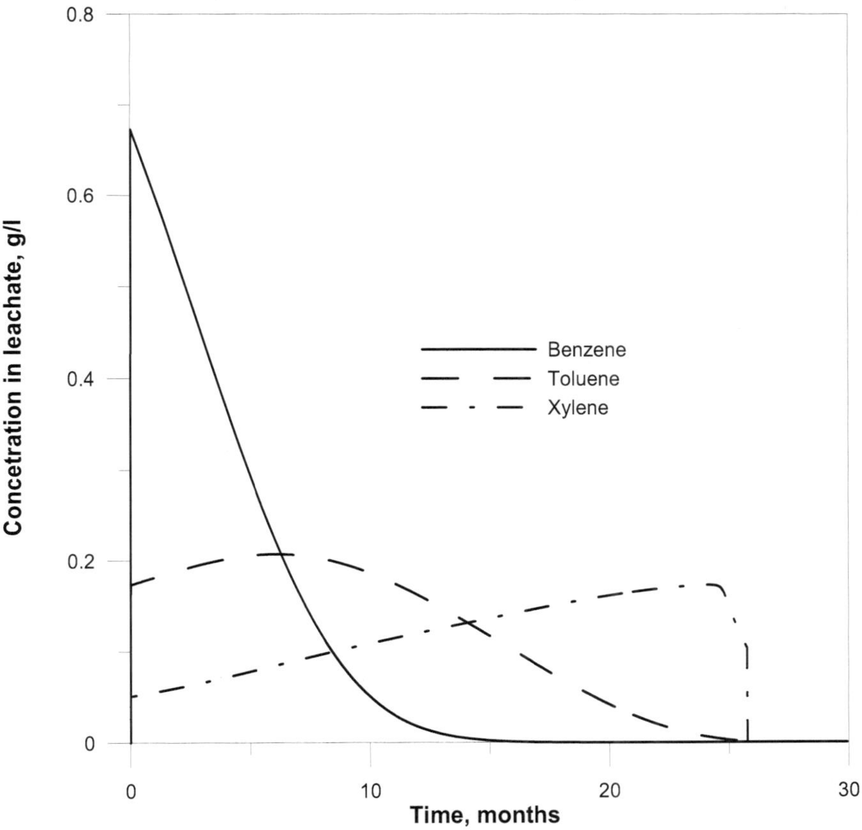

Figure 2: Aqueous concentration for different compounds in leachate.

role for eight months before xylene catches up and becomes the more significant pollutant. Since the three compounds have close affinity to the soil matrix, the variation in the compound concentration in the leachate could be explained by realizing that benzene has a solubility in water that is about three times as much as toluene and ten times compared to xylene. Examining the values from fig. 2, it is clear that the concentration levels are much higher than the MCLs and that as one concentration decreases the others take over which is expected to have a drastic effect on the underlying aquifer. The mass per unit area of the zone of contamination that eventually will reach the water table for the three compounds is shown in fig. 3. Notice that initially benzene is the main pollutant leveling in ten months. The contaminants concentration in groundwater will be obviously affected by this variable source by resulting in a pollution plume with variable characteristics making the remediation process more difficult. The leaching process will result in reducing the hydrocarbon in the vadose zone initially at a higher rate. However as the process continues the rate becomes much smaller

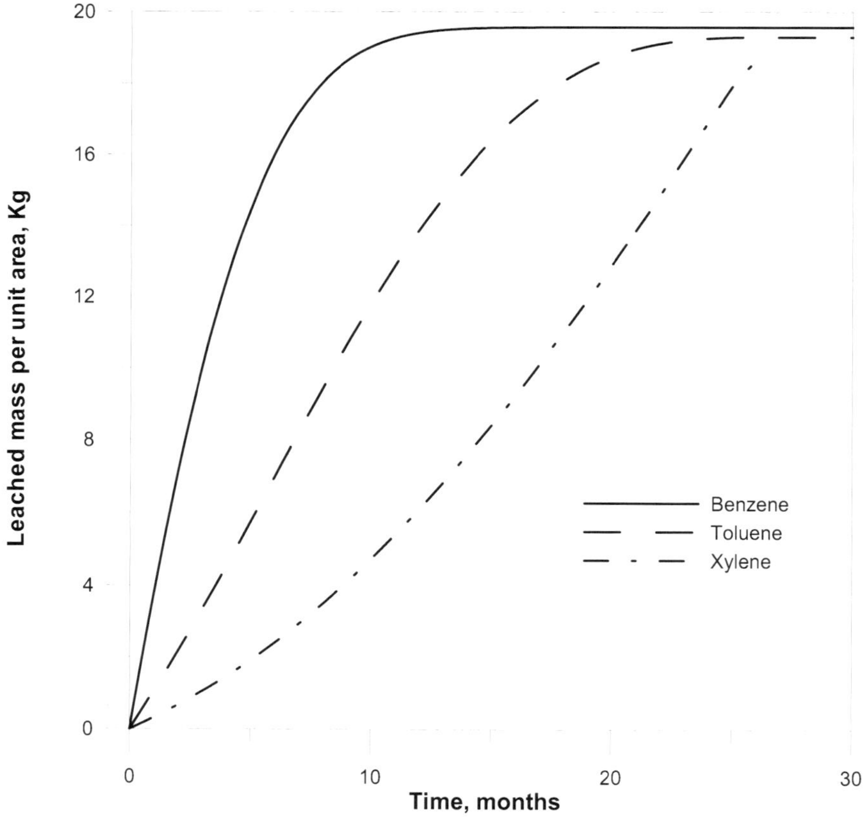

Figure 3: Total mass (pollution source) entering groundwater with the leachate.

(see fig. 4). The influence of the water flow rate is examined by plotting the residual hydrocarbon content for different rates as shown in fig. 4. Careful examination of the figure reveals that although higher water flow rate results in quicker removal of contaminants from the soil matrix, it takes it also transport the pollutants to groundwater faster. On the other hand lower water rate leads to slow but long acting pollution source.

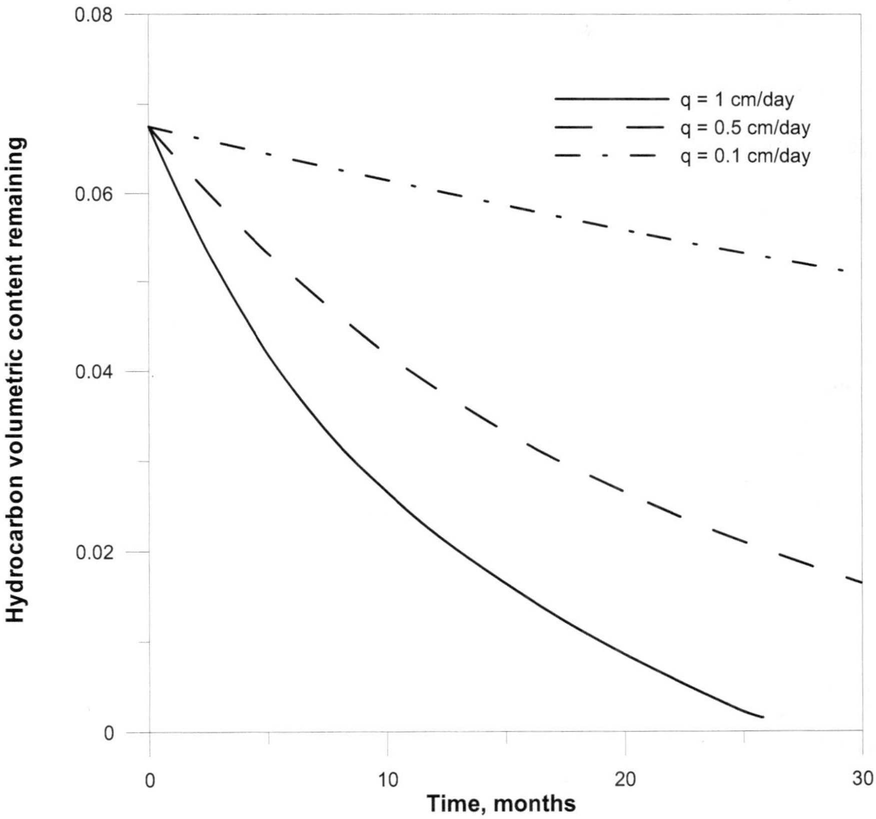

Figure 4: Hydrocarbon volumetric content remaining in vadose zone for different water flow rates.

The values used for this modeling study, in particular the water flow rate, might be very high, however it was selected in order to make assumptions like negligible contaminant degradation acceptable.

6 Status of fuel storage tanks in Saudi Arabia

In a recent study (Al-Suwaiyan et al. [9, 10]) a survey and site visits were conducted in order to evaluate the potential risk of groundwater pollution due to underground fuel storage tanks. Important information related the storage tanks, routine monitoring, maintenance, preventive maintenance as well as laws and regulation existence as well as enforcement were collected and reviewed. The results revealed that a significant number of service stations had old tanks made using reinforce concrete and that they never received maintenance which is a must given the aggressive subsurface environment. The study also showed the lack of protection from corrosion and only about 20% had cathodic protection. Any monitoring was absent in close to 60% of the service stations surveyed. Similarly, emergency response practices were lacking. Given the limited geographical area for this study, it still can be expected to roughly represent the actual conditions in many developing countries. This lack in all aspects, leads one to conclude that the risk for contaminating the underlying aquifers is high.

7 Summary and Conclusion

A compositional model was developed in order to simulate BTX concentrations in leachate from a subsurface contaminated by residual hydrocarbon. The model was also used to estimate the mass of contaminants that reach an underlying water table as well as quantifying the amounts remaining in the contaminated zone. The modeling results indicated the existence of very high concentrations of BTX in the leachate and it also revealed the long time effect of such pollution. The effect of the percolation rate was also studied to conclude that high flow rate will case quicker cleanup but higher concentration in groundwater and vice versa. Modeling is an essential part of remediation projects and any design or selection of operating condition and in turn the overall performance will be influenced by how accurately real conditions were presented in the model. This accounts for the fact that remediation of contaminated aquifers is rarely successful. Review of the status of underground fuel storage tanks at service stations in eastern Saudi Arabia could be extrapolated to some developing countries suggesting good potential for soil and groundwater pollution at many locations sooner or later. One way to avoid such problem could be through reducing or preventing, if possible, groundwater contamination in the first place by regulating underground fuel storage tanks and requiring continuous maintenance and monitoring.

Acknowledgement

The support provided by the Civil Engineering Department at King Fahd University of Petroleum and Minerals is highly appreciated.

References

[1] Van der Leeden, F., Troise, F.L. and D.K. Todd, (eds.). *The Water Encyclopedia*, 2nd ed., Lewis Publishers, Chelsea, MI, 1990.

[2] Al-Suwaiyan, M.S., Bashir, K., Aiban, S.A. and Ishaq, A.M., Analytical model to quantify crude oil spills in sandy layered aquifers. *Journal of Environmental Engineering*, ASCE, 128(4), pp. 320-326, 2002.

[3] Mishra, S., J. C. Parker, and N. Singhal, Estimation of soil hydraulic properties and their uncertainty from particle size distribution data. *Journal of Hydrology* 108, pp. 1-18, 1989.

[4] Saleem, M. Al-Suwaiyan, M., Aiban, S., Ishaq, A.M., Al-Malack, M. and M. Hussain. Estimation of spilled hydrocarbon volume-the state-of-the-art, *Environmental Technology*, 25(9), pp 1077-1090,2004.

[5] Farr, A.M., R.J. Houghtalen, and D.B. McWhorter, Volume estimation of light nonaqueous phase liquids in porous media, *Ground Water* 28(l), pp. 48-56,1990.

[6] Lenhard, R.J., and J.C. Parker, estimation of free hydrocarbon volume from fluid levels in monitoring wells, *Ground Water* 28(l), pp. 57-67, 1990.

[7] United States Environmental Protection Agency (USEPA). The Office of Ground Water & Drinking Water Web Site, Washington DC, http://www.epa.gov/OGWDW

[8] Charbeneau, R.J., Groundwater Hydraulics and Pollutant Transport, Upper Saddle River, New Jersey: Prentice –Hall, 2000.

[9] Al-Suwaiyan, M.S., A.M. Ishaq, M.H. Essa and M. Saleem, Assessment of the status of USTs at service stations in eastern Saudi Arabia. Proc. of the Fourth Saudi Technical Conference, Riyadh, pp. 286-293, 2006.

[10] Al-Suwaiyan M.S., A.M. Ishaq, M.H. Essa and M. Saleem, Environmental Pollution Due to Underground Fuel Storage Tanks: a)assessment; b) guidelines for prevention, KACST, Riyadh, 2008.

Acute toxicity of lead nitrate to red swamp crayfish, *Procambarus clarkii*

A. Balarezo & P. B. Tchounwou
Environmental Toxicology Research Laboratory,
NIH-RCMI Center for Environmental Health, College of Science,
Engineering and Technology, Jackson State University, USA

Abstract

Red crayfish, *Procambarus clarkii,* is widely farmed in the southern states of Louisiana and Texas. It is the most valuable of commercial crayfish species in the United States, where it is considered a delicacy. In recent years, its availability has decreased due to environmental contamination by toxic chemicals including lead. In the present study, we conducted a ninety-six-hour static renewal bioassay to assess the acute toxicity of lead as $Pb(NO_3)_2$ to adult red swamp crayfish. Study results indicated that $Pb(NO_3)_2$ is toxic to *P. clarkii,* and its toxicity is both time- and concentration-dependent. The 96-hr LC_{50} was computed to be 3.95 g/L. During experimentation, erratic behaviors such as restlessness, loss of balance, air gulping, and convulsion were observed in lead-exposed crayfish. Findings from this study have provided a scientific basis for designing subsequent experiments to assess the chronic exposure and biomarkers of lead-induced toxicity in *P. clarkii.*
Keywords: red swamp crayfish, lead nitrate, acute toxicity.

1 Introduction

Lead (Pb) is a ubiquitous metal that exists in several oxidation states (0, I, II, and IV). However, Pb^{2+} is the most stable, available, and suspected of being accumulated by aquatic organisms. The overall concentration of Pb in the continental crust is estimated to be 20 ppm dry matter. It is number 35 most abundant element in nature. Background levels in the top soil vary between 10 to 70 ppm; levels in surface water are generally bellow 0.01 ppm, but levels up to 1ppm can be expected inn contaminated areas with soft waters; in ocean water

WIT Transactions on Ecology and the Environment, Vol 132, © 2010 WIT Press
www.witpress.com, ISSN 1743-3541 (on-line)
doi:10.2495/ETOX100101

is about 0.0027 ppb. In U.S. sediments it is about 9-35 ppm, and about 4-700 ppm in soils [1–3].

Pb enters the aquatic environment through erosion and leaching from soil, dust fallout, combustion of fossil fuels, industrial water discharges, runoff and fallout deposits from streets and other surfaces as well as precipitation. Pb is known to accumulate in fish tissues, including bone, gills, kidneys, liver, and scales. Its toxicity is influenced by fish life stage, water pH, and hardness, and the presence of organic materials [4–7].

The use of Pb has resulted in increased levels in soil, water, and air; Pb atmospheric concentrations as high as 50 pg/m^3 has been found in industrialized regions [8]. In addition, areas next to mining activities may be exposed to high emission levels. As Pb particles (dust) may be transported via air, an individual source of emission may pollute areas located far away from its source. Industrial use of Pb may comprise mining, smelting, and processing, Pb- containing water pipes, plumbing solders, alloys, pigments, batteries, and ceramics and glassware [7, 8].

Historically, Pb in solders and alloys for water drinking pipes and as additive to gasoline have been the major source of environmental pollution, and animal and human exposure. In recognition of the toxic effects of Pb, most countries have phased out the use of Pb-E_4 in the fossil fuel energy industry. However, the emission of Pb from waste incinerators and waste disposals still remains. Accidental exposure of animals may also result from Pb shuts, disposed linoleum, and from Pb containing ornaments, toys, and pigments. Particularly, intoxications of animals resulting from disposed batteries ingestion has been reported [8].

Pb accumulation in soils and surface water depends on many factors including pH, mineral composition, and type and amount of organic material. Pb in soils is transferred to food crops. Roots usually contain more Pb than stems and leaves, while seeds and fruits contain the lowest concentrations [9].

Several studies have demonstrated that aquatic invertebrates exhibit various degrees of sensitivity to lead toxicity [10–17]. These invertebrates include the American red crayfish, *Procambarus clarkii*, which is native to the Louisiana marshes (USA) [18, 19]. It is commercially farmed and harvested as a very important food source, both to fishing industries and to recreational and subsistence fishermen in Louisiana and other southern states [20]. Additionally, it has been reported that crayfish are being fished commercially for consumption without adequate protection to human health [12]. They constitute a commercially valuable natural renewable resource [21]. Also, they live in a wide range of environmental conditions that include highly polluted waters resulting in high resistance to heavy metals [13, 18, 22]. Hence, it is important to investigate the sensitivity of this valuable ecologic resource to toxic chemicals. The present study was designed to determine the acute toxicity of lead to red swamp crayfish in order to gather the necessary scientific information for the conduct of subsequent chronic experiments.

2 Materials and methods

Test organisms (mean body weight 15.3-28.5 g) were obtained from a local business supplier in Jackson, MS. They were initially declawed to preclude cannibalism and predation. Prior to experimentation, specimens were acclimatized at laboratory conditions for 2 weeks period, into several 50 gallon glass aquaria filled with 20 gallons of dechlorinated tap water. Crayfish were fed sinking waffles fish food *ad libitum* once every other day to avoid water fouling and reduce feces generation. Feeding was stopped 24 hr prior to testing and during testing.

Ninety-six hours static renewal bioassay was conducted according to standard testing protocols. Range-finding, subsequent and definitive experiments were conducted according to 8910 American Public Health Association (APHA) standardized procedures [23]. A completely randomized design with 6 treatments [0, 2, 4, 6, 8, and 10 g/L $Pb(NO_3)_2$] was deployed with three replicates per concentration and 4 organisms in each replicate. Tests were carried out in 2-gallon glass aquaria filled with 4 liters of Pb-solution at different concentrations. Four randomly sorted crayfish were placed in each of the 6 test chambers and their replicates. The bioassays included a total of 18 experimental units carried out during 96-h period. The tests were repeated two times to ensure reproducibility. No-aeration, no-feeding, but daily renewal of freshly prepared and aerated solutions were maintained during the 96-h testing period.

Lethality was observed every 2 hr for the first 12 hours of exposure, and then after every 24 hr. Dead crayfish were immediately removed to keep water quality. The percentages of mortality or viability at each 24 hr of exposure were calculated. To determine the LC_{50}, a linear correlation between crayfish mortality and decimal logarithm of Pb-concentration was established; the correlation equation and regression coefficient, R^2, were obtained.

During experimentation, basic water quality parameters including temperature, pH, dissolved oxygen, hardness and alkalinity were analyzed following standard protocols [23]. A photoperiod of 11h light and 13h darkness was maintained.

3 Results

The physicochemical characteristics of the laboratory water, supplied for all the bioassays carried out in this research, were in the following range: pH = 6.8-7.2; dissolved oxygen = 7.5-8.3 mg/L; temperature: 18-22°C, alkalinity 10-20 mg/L as $CaCO_3$, and hardness 12-15 mg/L as $CaCO_3$.

Acute exposure to lethal concentrations of Pb resulted in noticeable impact on crayfish viability. The mean values of viability of crayfish exposed to various concentrations of Pb^{2+} for 24, 48, 72, and 96 hr are presented in Figure 1. It was observed that during the exposure period, all crayfish in the control group survived. Within the first 24 hr of exposure, no crayfish mortality was observed for all Pb-concentrations (2, 4, 6, 8, and 10 g/L) assayed. After 48 hr of exposure the percentage of crayfish mortality markedly increased from 25 to 87%. After 72 hr of exposure, 100% crayfish mortality was observed for the three higher Pb-concentrations (6, 8, and 10 g/L) . After 96 hr exposure, viability was only noticed for the 2 g/L Pb-bioassays. In addition, results demonstrated that viability of crayfish and concentration of Pb in the exposure medium were negatively related; in other words, mortality rates of the crayfish increased with increasing in Pb-concentration and exposure period.

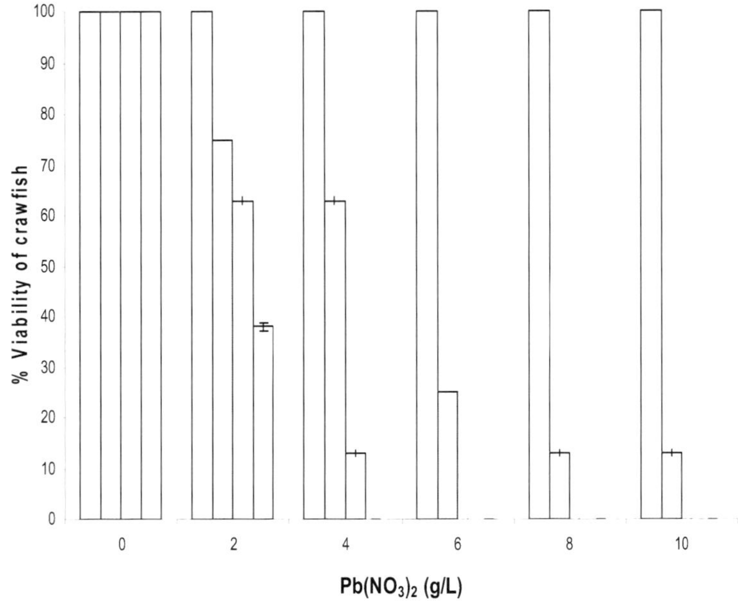

Figure 1: Effect of $Pb(NO_3)_2$ on the viability of adult *Procambarus clarkii* at 24, 48, 72, and 96 hr of exposure.

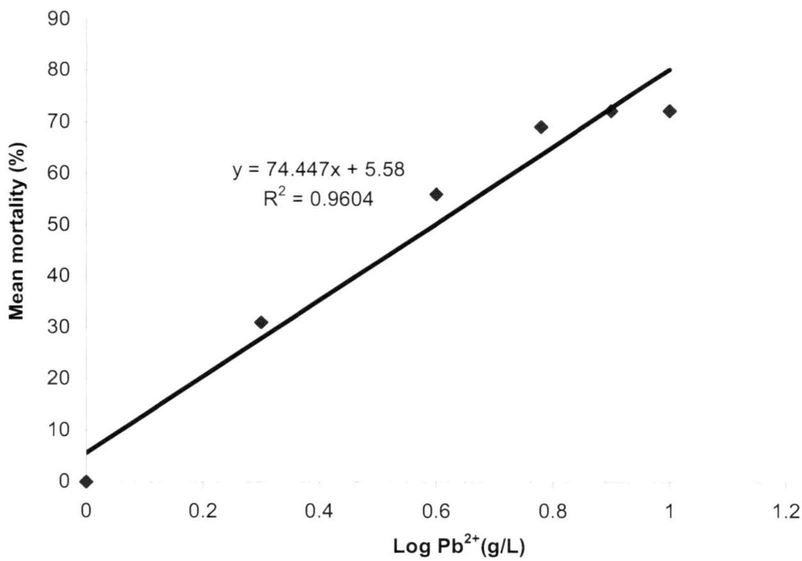

Figure 2: Correlation of crayfish mortality (%) and Pb-concentration (g/L) at 96-hr exposure.

During experimentation, lead-exposed crayfish exhibited various behavioral patterns before death occurred; restlessness, motionless, loss of balance (i.e. over turning), air gulping and convulsion were frequently observed. However, such responses were minimal in the groups exposed to 2.0 g/L $Pb(NO_3)_2$ and none at all in controls.

Clearly, both time- and concentration-response relationships were observed in these bioassays. Figure 2 shows the regression analysis of the relationship between crayfish mortality and $Pb(NO_3)_2$ concentration after 96 hr of exposure. From the regression curve, a 96-hr LC_{50} value was calculated to be 3.95 g/L.

4 Discussion

This study demonstrated that lead, as $Pb(NO_3)_2$ is toxic to crayfish, and its toxicity in time- and concentration-dependent. Upon 96 hr of exposure the LC_{50} was 3.95 g/L, which is considerably higher than the 751.57 mg/L reported by Naqvi and Howell [15] from their investigation of the effect of lead nitrate on the fecundity of juvenile Louisiana swamp crayfish (*Procambarus clarkii*). The major possible reasons for this marked difference in values of LC_{50} could be (a) organisms source and age: laboratory-raised versus farm-raised crayfish, and juvenile versus adult crayfish; (b) water quality of bioassays: aged tap versus synthetic. The water quality used in our research, although not aged tap, was acceptable according to EPA-807-8720 (Toxicity test procedures for crustaceans); and (c) the actual Pb-concentration derived from nitrate salt [100 mg/L as $Pb(NO_3)_2$ = 16.84 mg/L Pb] versus 60.72 mg/L of Pb for the same $Pb(NO_3)_2$ concentration. Age and size of specimens have been reported to influence to the extent of masking observed trends of bioconcentration [24]. In addition, in the present research 100% crayfish viability was observed within the 24 hr exposure to 2 g/L, compared to 8% mortality at 1 g/L $Pb(NO_3)_2$ reported by Naqvi and Howell [15].

Erratic behavior such as restlessness, loss of balance, air gulping, and convulsion observed during experimentation are in agreement with earlier reports by Anderson [10] and Naqvi and Howell [14, 15]. These behavioral responses are indications of toxicity due to nervous disorders and insufficient gaseous exchange across the gill epithelia they mentioned. The toxic manifestation would be consistent with the inhibition of the enzyme acetyl-cholinesterase, resulting in death by paralysis of muscles of respiration and/or depression of the respiratory center [25]. Other studies have reported that that Pb is more toxic at low pH [8, 26].

Acknowledgements

This research was financially supported in part by a grant from the National Institutes of Health (No. 2G12RR013459), and in part by a grant from the National Oceanic and Atmospheric Administration – NOAA Grant No. NA17AE1626, Subcontract No. 27-0629-017 to Jackson State University.

References

[1] Galvin R.M. Occurrence of metals in water: An overview. *Water SA*, **22,** pp. 7-18, 1996.
[2] WHO-IPCS. Inorganic lead, Environmental Health Criteria 165. World Health Organization, Geneva, Switzerland. 1995.
[3] Tchounwou P.B. A.A. Siddig & M.L. Marian. Post-remediation monitoring for soil, sediment, and water contamination by lead from a controlled superfund site in Mississippi. *In: Environmental Toxicology and Risk Assessment: Eighth Volume, ASTM STP 1381*. F.T. Price, K.V. Brix, and N.K. Lane, Eds. American Society for Testing and Materials, West Conshohocken, PA. pp. 16-32, 2000.
[4] Hellawell J.M. *Biological Indicators of Freshwater Pollution and Environmental Management.* Elsevier Applied Science Publishers. London. 1986.
[5] Moore J.W.W. & S. Ramamoorthy. Heavy metals in natural waters, In: *Applied Monitoring and Impact Assessment*. Springer Verlag, N.Y., 1984.
[6] Merlini M. & G. Pozzi. Lead and fresh water fishes: Part 2-Ionic lead accumulation. *Environ. Pollut*. **13**, pp. 119-126, 1997.
[7] WHO-IPCS. *Lead, Environmental Criteria 85*. World Health Organization, Geneva, Switzerland. 1989.
[8] ATSDR (Agency for Toxic Substances and Diseases Registry). *Toxicological Profile for Lead (Update)*. Center for the Disease Control and Prevention, Agency for Toxic Substances and Disease Registry, Atlanta, GA. 1999.
[9] Davies D.J., J.M. Watt, & I. Thornton. Lead levels in Birmingham dust and soils. *Sci. Total. Environ*. **67**, pp. 177-185, 1987.
[10] Anderson R.V. The effect of lead on oxygen uptake in the crayfish, Orconectes virilis (HAGEN). *Bull. Environ. Contam. Toxicol*. **2,** pp. 394-400, 1078.
[11] Tulasi S. J., R. Yasmeen, C. P. Reddy & J.V. R. Rao. Lead uptake and lead loss in the freshwater field crab, Barytelphusa guerini, on exposure to organic and inorganic lead, *Bull. Environ. Contam. Toxicol*. **39,** pp. 63-68, 1987.
[12] Pastor A., J. Medina, J. Del Ramo, A. Torreblanca, J. Diaz-Mayans & F. Hernandez. Determination of lead in treated crayfish Procambarus clarkii: accumulation in different tissues. *Bull. Environ. Contam. Toxicol*. **41**, pp. 412-418, 1988.
[13] Martinez M., A. Torreblanca, J. Del Ramo, & J. Diaz-Mayans. Effects of sublethal exposure of lead on levels of energetic compounds in Procambarus clarkii (Girad, 1852), *Bull. Environ. Contam. Toxicol*. **52**, pp. 729-733, 1994.
[14] Naqvi, S.M. & R.D. Howell. Cadmium and lead uptake by red swamp crayfish (Procambarus clarkii) of Louisiana. *Bull. Environ. Contam. Toxicol*. **51**, pp. 296-302, 1993.

[15] Naqvi S. M. & R. D. Howell. Toxicity of cadmium and lead to juvenile red swamp crayfish (Procambarus clarkii), and effects on fecundity of adults. *Bull. Environ. Contam. Toxicol.* **51**, pp. 303-308, 1993.
[16] Devi M. & M. Fingerman. Inhibition of acetylcholinesterase activity in the central nervous system of the red swamp crayfish, Procambarus clarkii by mercury, cadmium, and lead. *Bull. Environ. Contam. Toxicol.* **55,** pp. 746-750, 1995.
[17] Tungare S. M. & A. D. Sawant. Lead and cadmium in selected species of shrimp around the Mumbai coast, India. *Bull. Environ. Contam. Toxicol.* **68**, pp. 455-462, 2002.
[18] Abdelghani A. A., Y.V. Pramar, T.K. Mandal, P.B. Tchounwou, & L. R. Heyer,. Levels and toxicities of selected inorganic and organic contaminants in a swamp environment. *J. Environ. Sci. & Health*, **B30 (5),** pp. 717-731, 1995.
[19] Torreblanca A., J. Del Ramo, & J. Diaz-Mayans. Effects of cadmium on the biochemical composition of freshwater crayfish Procambarus clarkii, (Girard, 1852). *Bull. Environ. Contam. Toxicol.* **47,** pp. 933-938, 1991.
[20] Green R.M. & A. Abdelghani. 2004. Toxicity of a 2,4-dichorophenoxyacetic acid and monosodium methanearsonate to the red swamp crayfish, Procambarus clarkii. *Int. J. Environ. Res. Public Health* **1**, pp. 35-38, 2004.
[21] Naqvi S. M. I. Devalraju, & N. H. Naqvi. 1998. Copper bioaccumulation by red swamp crayfish, Procambarus clarkii, *Bull. Environ. Contam. Toxicol.* **61**, pp. 65-71, 1998.
[22] Torreblanca A., J. Del Ramo & J. Diaz-Mayans. 1989. Gill ATPase activity in Procambarus clarkii as an indicator of heavy metal pollution. *Bull. Environ. Contam. Toxicol.* **42**, pp. 829-834, 1989.
[23] APHA (American Public Health Association). Standard Methods for Examination of Water and Wastewater. Editors: S. Clesceri, A. E. Greenberg & A. D. Eaton, 20th Edition, Washington, DC. 1998.
[24] Vinikour W.S. R.M. Goldstein & R.V. Anderson. Bioaccumulation patterns of Zn, Cu, Cd, and Pb in selected fish species from the Fox River, Illinois. *Bull. Environ. Contam. Toxicol.*, **24**, pp. 727-734, 1980.
[25] Cearly J.E. & R.L. Coleman. Cadmium toxicity in largemouth bass and bluegill. *Bull. Environ. Contam. Toxicol.* **11**, pp.146, 1974.
[26] Schubawer-Berigan M.K., J.R. Dicken, P.D. Monson, & G.T. Ankly. pH-dependent toxicity of Cd, Cu, Ni, Pb and Zn to Ceriodaphnia dubia, pimaphales promelas, Hyalella azteca, and Lumbriculus variegates. *Environ. Toxicol. Chem.* **12,** pp. 1261-1266, 1993.

Section 3
Biodegradation, bioremediation and biomonitoring

Biostimulation combined treatments for remediation of diesel contaminated soil

C. Calvo[1], G. A. Silva-Castro[1], I. Uad[1], M. Manzanera[1],
C. Perucha[2], J. Laguna[2] & J. Gónzalez-López[1]
[1]Department of Microbiology, Institute of Water Research, University of Granada, Spain
[2]AG Ambiental, Spain

Abstract

Bioremediation is an important technology for the restoration of oil polluted environments by indigenous or selected microorganisms. In general, the rate of biodegradation depends on the number and types of microorganisms, the nature and chemical structure of pollutants to be degraded and the environmental conditions. In this study we have evaluated the efficacy of the application of four different biostimulation treatments for the biodegradation of diesel contaminated soils. The treatments applied involved: (a) the addition of NPK fertilizer + Ivey I surfactant; (b) the addition of NPK fertilizer + Ethanol; (c) the addition of NPK fertilizer + Biorem; and (d) oxidation by Fenton's reagent combined with NPK fertilizer. Microbial activity was evaluated following growth of heterotrophic and degrading microorganisms, dehydrogenase activity and CO_2 production. Hydrocarbons degradation was established by determination of TPH, alkanes, branched alkanes, pristane and phytane by GC/MS. Our results have shown that the application of NPK fertilizer in combination with Ivey surfactant is an efficient treatment to be applied in clay soil. Treatment with Fenton's reagent previous to the application of NPK fertilizer also efficiently enhanced hydrocarbon biodegradation in saturated conditions.
Keywords: bioremediation, hydrocarbon pollution, surfactant, Fenton reagent.

1 Introduction

Bioremediation offers a more environmentally friendly alternative by taking advantage of the oil degrading microorganisms and by establishing and

doi:10.2495/ETOX100111

maintaining the physical, chemical and biological conditions that favour enhanced oil biodegradation rates in the polluted environment [1]. Biological processes have been used successfully to remediate soils polluted with petroleum hydrocarbons and their derivatives [2]. It is considered an environmentally acceptable way of eliminating oils and fuel because the majority of hydrocarbons in crude oils and refined products are biodegradable, and hydrocarbons degrading microbes are ubiquitous [3]. One of the best approaches to restore polluted soils consists in using microorganisms able to degrade those toxic compounds for bioremediation processes [4].

Biostimulation process introduces additional nutrients into a polluted system to increase indigenous microorganisms. Contamination of a zone with hydrocarbons originates a rapid depletion of the available pools of major inorganic nutrients, such as N and P. Consequently, nutrient supplementation for hydrocarbons degradation has been traditionally focused on addition of nitrogen and phosphorous either in the organic or inorganic forms [5].

Besides adding nutrients to accelerate the breakdown of oil by microorganisms, another factor that enhances oil biodegradation consists in enlarging oil dispersion addition of either by chemical or biological surfactants. In general, microbial attack takes place at the oil-water interface, thus enhanced biodegradation should results as a consequence of the increasing surface area available for microbial colonization [6].

Emulsifiers can emulsify hydrocarbons by enhancing their water solubility and increasing the displacement of oily substances from soil particles [7]. For these reasons, inclusion of surfactants in a bioremediation treatment of a hydrocarbon polluted environment could be of great advantage. In consequence, addition of surfactant and other natural emulsifying agents are important tools for biotreatment of hydrocarbon polluted environment [8].

On the other hand, bioremediation is often limited to those compounds that are biodegradable, and not all compounds are susceptible to rapid and complete degradation [9], for this reason, application of combined processes based on combination between oxidation and biological treatment have been used for the removal of some pollutants present in soil. Among the different available oxidants, hydrogen peroxide and Fenton's reagent have been widely and successfully applied for the remediation of many organic compounds [10]. Fenton reaction causes the dissociation of the oxidant and the formation of highly reactive hydroxyl radicals. The unstable hydroxyl radicals formed are used to degrade organic compounds either by hydrogen abstraction or by hydroxyl addition [11]. This reaction increases availability source of carbon to microorganisms in biological treatment. They can also be combined with bioremediation to design complete soil treatment processes [12].

Biological degradation represents one of the major routes through which hydrocarbons can be removed from polluted environments. To show the potential bioremediation technology, it is important use of to enhance the rate of hydrocarbon biodegradation under controlled conditions. The laboratory feasibility study involves microbiological and chemical methods to measure the effectiveness of bioremediation under predetermined conditions.

In this study, laboratory microcosms were carried out to measure the changes in microbial activities and hydrocarbons biodegradation during bioremediation treatment. The aim of this research was to understand the extent of oil hydrocarbon degradation under different experimental treatments combining inoculation with NPK fertilizer and Ivey surfactant, Biorem, ethanol as biostimulating agents and pre-treatment with Fenton reagent as oxidant agent.

2 Materials and methods

2.1 Soil samples

In this study we used was loam clay soil polluted at the origin and supplied by AG Ambiental S.L. This soil was composed of 36% clay, 33% sand, 1.67 COT, 577 mg/kg Nitrogen, 299mg/kg Phosphorus and 28135 mg/kg Iron.

2.2 Biostimulating agents

NPK inorganic fertilizer (18:8:17 Agroblem SA) was composed of 18% total Nitrogen; 8% Phosphorus pentoxide (P_2O_5); 17% Potassium oxide (K_2O); 2% Magnesium oxide (MgO) and 19% Sulphur trioxide (SO3). Ivey-sol® Surfactant comprised of several patented and preparatory non-ionic surfactant formulations.

Biorem combination is an organic fertilizer composed of 31.5% COT, 3% Nitrogen, 0.06% Phosphorus and 1% Potassium. Ethanol is classified as a primary alcohol, a group of chemical compounds whose molecules contain a hydroxyl group, bonded to a carbon atom.

2.3 Oxidation agents

Fenton's reagent is a mixture of H_2O_2 and ferrous iron, which generates hydroxyl radicals. The factors more important in Fenton process are the relationship between Oxidant agent (H_2O_2) and Organic compounds (RH) and the pH. The optimum pH is around 3.

2.4 Microcosm assays

Microcosms were built in 500 ml Erlenmeyer flask, each flask containing 250 g of soil. Experiments were prepared with the NPK fertilizers as the principal biostimulation agent, supplemented or not with other biostimulation additive i.e. Ivey surfactant, Biorem, ethanol and Fenton's reagent [11].

All supplemented additives were applied at two different concentrations and set up at room temperature. Treatments are detailed in table 1.

2.5 Enumeration of culturable bacteria in soil

Three replicate samples from each microcosm treatment were withdrawn every week for enumeration of aerobic heterotrophic bacteria and degrading bacteria. 0.1 ml of serially diluted soil samples were plated on 1/10 diluted Trypticase Soy

Table 1: Experimental design.

	Treatment	*NPK fertilizer*	*Ivey Surfact*	*Biorem*	*Ethanol*	*Fenton*
A	Natural attenuation	-	-	-	-	-
B	NPK	0.3 g/ kg of soil	-	-	-	-
C	NPK + Ivey® Surfactant	0.3 g/ kg of soil	52 μl/Kg	-	-	-
D	NPK + Biorem	0.3 g/ kg of soil	-	5g/kg	-	-
E	NPK + Ethanol	0.3 g/ kg of soil	-	-	0.2%w/w	-
F	[H_2O_2], Fe^{+3}+ NPK	0.3 g/ kg of soil	-	-	-	11.5 ml

Agar (TSA, Difco). Degrading bacteria were counted on 1% hydrocarbon trypticase soy agar. Triplicate plates were incubated at 28°C for 48 h before the colonies were counted.

2.6 Biological activity

Dehydrogenase activity was determined by the reduction 2,3,5-tryphenylterazolium chloride (TTC) to tryphenyl formazan (TPF) according [13].

Production of CO_2 was determined by gas chromatography (Varian Star 3400 cx, with TCD Detector). Soil samples for the respiration test, were incubated in a closed vessel at 28°C [14].

2.7 Hydrocarbon analysis

Total petroleum hydrocarbons were determined using EPA 8015 (GROs + DROs) [15]. The gasoline range organics (GROs) were introduced into the GC/FID by purge and-trap, automated headspace, and the Diesel range organics (DROs) were prepared by soxhlet extraction .

Analysis of each hydrocarbon fraction were performed from the extracted fractions above mentioned using a Hewlett–Packard 6890 GC system equipped with a HP-5-MS-capillary column (30 m×0.32 mm I.D. Fraction of hydrocarbons were detected using a mass detector 5872 (Hewlett–Packard) and a library utilized was Wiley 275.

2.8 Statistical analyses

Mean, variance and standard deviation of the microbiological and chemical parameters were calculated from the values obtained in each measurement of the triplicate samples. Differences between biological and chemical analysis in the different soil samples were tested by student test. The statistical significance was evaluated at the $P < 0.05$. All statistical analyses were carried out using the SPSS 15.0 software.

3 Results and discussion

Microcosm test can be used to assess the biodegradation potential of hydrocarbons. Screening of bioremediation treatments can be used to design the most appropriate bioremediation strategy for large scale application. In this study, we have compared the effect of various bioremediation treatments on hydrocarbon removal. Strong correlation between microbial counts and hydrocarbons degradation has been reported by Al-Awadhi et al. [16]. The results obtained in this study have shown that Surfactant (C) and Fenton (F) treatments increased the number of both heterotrophic and hydrocarbon degrading bacteria (figure 1). In addition, the number of bacteria detected at the beginning of treatment (near 10^7) was upper than those usually counted in non polluted soil, indicating that polluted samples studied have been enriched in hydrocarbon tolerant microorganisms. These results could be explained by an adaptation of microbiological population to the pollutants. Also, none of the nutrients added inhibit microbial growth, indicating that no toxic effect was observed on microbial populations.

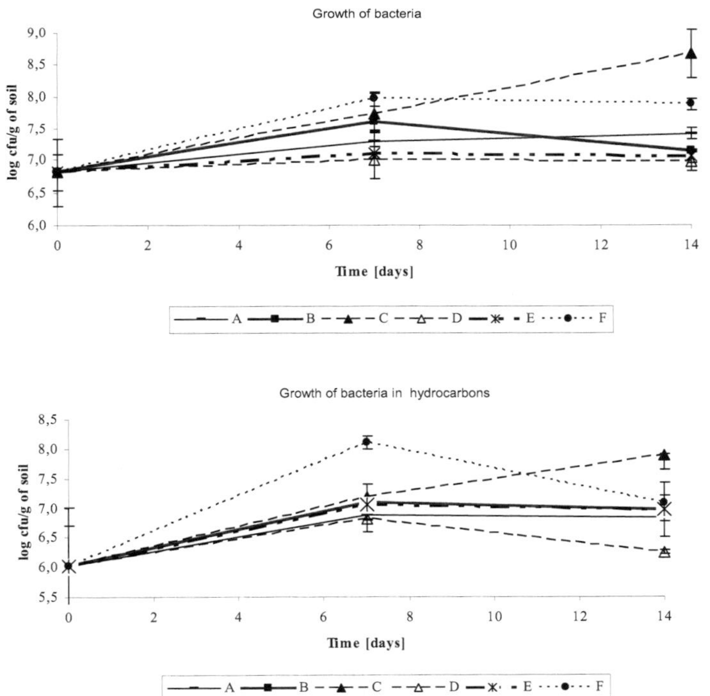

Figure 1: Number of total heterotrophic and degrading bacteria during treatment.

Dehydrogenase activity measured in soil has been used to monitor activity as an index for the total oxidative activity of microorganisms. In general, biological oxidation of organic compounds is generally a dehydrogenation process, which is catalyzed by dehydrogenase enzymes [17]. Table 2 shows data of biological activity obtained at the beginning and end of soil treatment. We observed high dehydrogenase activity at the beginning of treatment B, C with 429.7 TPF/g, and 411.1 TPF/g respectively. High values of dehydrogenase activity were correlated with stimulation of microbial populations measured at the end of the experiments.

Soil respiration is another measure of the total biological activity and result of degradation of organic matter (Table 2). Production of CO_2 decreased after 14 days of treatment, and only treatment D produced an increase in CO_2 production of 0.04% from 0.27% ($p<0.05$). These parameters showed a significant biological activity in these soil samples polluted at source meaning that microbial populations were already adapted to oil hydrocarbons pollution.

Table 2: Microbial activity estimated by moisture, dehydrogenase and production of CO_2.

Treat	Moisture		Dehydrogenase		%CO_2 24h	
	$t_{0\ days}$	$t_{14\ days}$	$t_{0\ days}$	$t_{14\ days}$	$t_{0\ days}$	$t_{14\ days}$
A	24,8±0,0	24,4±0,1	90,0±12,2	69,4±4,7	0,14±0,01	0,16±0,03
B	25,9±1,0	19,0±1,7	429,7±112,5	301,1±45,8	0,38±0,04	0,21±0,03[a]
C	34,2±0,8	33,5±1,0	411,1±131,3	371,5±37,1	0,35±0,05	0,17±0,03[a]
D	28,2±0,6	27,8±0,1	13,6±0,4	74,8±21,3[a]	0,04±0.00	0,21±0,04[a]
E	29,8±0,1	24,6±2,2	21,9±0,3	69,4±13,3[a]	0,19±0,01	0,19±0,02
F	34,6±0,5	38,4±2,6	106,9±30,83	120,3±7,92	0,18±0,0	0,16±0,0
P^a	*	*	*	*	*	*

P^a from one-way ANOVA, ***, **, * <0.001, 0.01, 0.5, respectively.
Labels (a) Indicates statistically significant between t=0 days and T=14 days, using t-student; P<0.05.

Chromatographic analysis was used to estimate the degradation of TPH, n-alkanes, branched alkanes, pristine and phytane as biomarkers. The effect of the bioremediation treatments on the degradation of hydrocarbon fractions is shown in Table 3. When comparing bioremediation treatments with natural attenuation, all treatments showed better rate of degradation in all fractions. However, its efficacy varied with the hydrocarbons fractions to be considered, thus for TPH removal, treatment B (NPK) was the most efficient (87.6%); but treatment C (NPK + surfactant) improved the removal of alkanes (94.1%). Finally, treatment with Fenton reagent achieved 100% of Pristane degradation.

Biological processes have traditionally been considered incompatible with chemical oxidation because of excessive death and inactivation of the native microorganisms [18]. However, Fenton's reaction may be recommended as a pre-treatment bioremediation process, because oxidation process reduces TPH,

Pristane and phytane. That oxidation does not affect microbial number and chemical and biological oxidation of contaminants in soil and can proceed simultaneously with treatment remediation. Mater et al. [19] suggested that the Fenton's reagent increased degradation of hydrocarbons in water bodies and wastewaters showing high efficiency when conditions are optimized.

In summary, treatability studies indicate that addition of nutrients and surfactants increase the rate of hydrocarbons biodegradation in soil. However, our results showed that addition of inorganic fertilizers after a pre-treatment with Fenton's reagent could be efficient treatment in the bioremediation of oil-contaminated soil, in particular in soils where the microbiological populations are adapted to the pollutants.

Table 3: Percentages and correlation indices of degradation in soil after biostimulating treatment.

	TPH		**Alkanes**		**Branched alkanes**		**Pristane**		**Phytane**	
Treat	%	Ic	%	Ic	%	Ic	%	Ic	%	Ic
A	19,6± 0,21	1	25,9±0, 20	1	39,9±0,16	1	39,6±0,1 6	1	40,0±0,16	1
B	87,6± 0,55	4,5	87,8±0, 54	4,5	88,9±0,66	4,5	92,1±0,9 7	4,7	91,4±1,05	4,6
C	80,9± 3,78	4,1	94,1±1, 14	4,8	72,9±5,20	3,7	72,7±5,2 3	3,7	73,7±5,06	3,7
D	54,6± 0,22	2,8	67,1±0, 16	3,4	56,5±0,21	2,8	60,1±0,1 9	3,1	62,1±0,18	3,2
E	70,3± 0,35	3,6	74,2±0, 30	3,8	75,9±0,28	3,8	77,9±0,2 5	3,9	78,4±0,25	3,9
F	70,1± 1,29	3,6	25,8±0, 44	1,3	48,0±5,95	2,4	100,0±0, 00	5,1	70,5±0,18	3,6
***P*[a]**	***		***		***		***		***	

***P*[a]** from one-way ANOVA, ***, **, * <0.001, 0.01, 0.5, respectively.
Ic: Correlation indices of degradation were calculated by the following expression: [% of degradation treatment/% of degradation Control].

Acknowledgement

This research has been supported by projects of Ministerio de Medio Ambiente (MMA. A4872007/20-01.1).

References

[1] Zhu, X., Venosa, A.D., Suidan, M.T., Lee, K., 2001. Guidelines for the bioremediation of Marine Shorelines and Freshwater Wetlands. Us Environmental Protection Agency.

[2] Mulligan, C.N., Yong and R.N., Gibbs B.F. 2001. Surfactant-enhanced remediation of contaminated soil: a review. Engineering Geology, 60, 371-380.

[3] Kanaly, R.A., Harayama, S. (2000). Biodegradation of high-molecular-weight polycyclic aromatic hydrocarbons by bacteria. Journal Bacteriology, 182, 2059-2067.

[4] Bento, F.M., Camargo, F., Okeke B., Frankenberger W. 2005.Comparative bioremediation of soils contaminated with diesel oil by natural attenuation, biostimulation and bioaugmentation. Bioresource Technology 96: 1049-1055.

[5] Sarkar, D., Ferguson, M., Datta, R., Birnbaum, S. 2005. Bioremediation of petroleum hydrocarbons in contaminated soil: Comparison, and monitored natural attenuation. Environmental Pollution 136, 187-195.

[6] Nikolopoulou, N., Kalogerakis, N., 2008. Enhanced bioremediation of crude oil utilizing lipophilic fertilizers combined with biosurfactants and molasses. Marine Pollution Bulletin, 56; 1855-1861

[7] Banat, R.S., Makkar, S.S., Cameotra. 2000. Potential commercial applications of microbial; surfactants, Applied Microbiology Biotechnology. 53: 495–508.

[8] Ron and E. Rosenberg. 2002. Biosurfactants and oil bioremediation, Curr. Opin. Biotechnol. 13, 249–252.

[9] Vidali. M. 2001. Bioremediation. An overview. Pure Appll. Chem., 73, 1163-1172

[10] Ferrarse, E., Andreottola, G. 2008. Application of advanced oxidation processes and electroxidation for the remediation of river sediments contaminated by PAHs. Journal of Environmental Science and Health Part A. 43, 1361-1372.

[11] Gan, S., Lau, E.V., Ng, N.K. 2009. Remediation of soil contaminated with polycyclic aromatic hydrocarbons (PAHs). Journal of Hazardous Materials. 172, 532-549.

[12] Valderrama, C., Alessandri, R., Aunola, T., Cortina, J.L., Gamisans, X., Tuhkanen, T. 2009. Oxidation by Fenton's reagent with biological treatment applied to a creosote-contaminated soil. Journal of Hazardous Materials. 166, 594-602.

[13] Tabatabai, M.A. (1982). Soil Enzymes. In: Page, A.L., Miller, R.H., Keeney, D.R. (Eds), Methods of Soil Analysis: Part 2. Chemical and Microbiological Properties. (pp 903-943) American Society of Agronomy, Soil Science Society of America, Madison, Wisconsin USA

[14] Bremner, M.J., Blackmer, A.M. (1982). Composition of Soil Atmospheres. In: Page, A.L., Miller, R.H., Keeney, D.R. (Eds), Methods of Soil Analysis: Part 2. Chemical and Microbiological Properties. (pp 903-943) American Society of Agronomy, Soil Science Society of America, Madison, Wisconsin USA.

[15] US Environmental Protection Agency, 1996.Hazardous waste- test Methods, Nonhalogenated organics by GAS Chromatography. 8015.

[16] Al-Awadhi, N., Al-Daher, R., El-Nawawy, A., Balba, M.T. 1996. Bioremediation of oil-contaminated soil in Kuwait I. Landfarming to remediate oil-contaminated soil. Journal of Soil Contamination, 5; 243-260.
[17] Paul, E.A., Clark, F.E., 1989. Soil Microbiology and Biochemistry, Academic Press, New York, 46-48.
[18] Ndjou'ou, A., Cassidy, D. 2006. Surfactant production accompanying the modified Fenton Oxidation of hydrocarbons in soil. Chemosphere 65, 1610-1615.
[19] Mater, L., Rosa, E.V.C., Berto, A.X.R., Corrêa, Schiwingel, P.R., Radetski, C.M. 2007. A simple methodology to evaluate influence of H2O2 and Fe2+ concentrations on the mineralization and biodegradability of organic compounds in water and soil contaminated with petroleum. Journal of Hazardous Materials 149, 379-386.

New isolation method of desiccation-tolerant microorganisms for the bioremediation of arid and semiarid soils

M. Manzanera[1], J. J. Narváez-Reinaldo[1,2], L. SantaCruz-Calvo[1], J. I. Vílchez[1,2], J. González-López[1,2] & C. Calvo[1,2]
[1]*Institute of Water Research, University of Granada, Spain*
[2]*Department of Microbiology, University of Granada, Spain*

Abstract

In arid and semiarid regions, the establishment of microorganisms and plants for the bioremediation of polycyclic aromatic hydrocarbons (PAHs) is further impeded by a number of physicochemical factors including low precipitation, high evaporation, high winds and extreme temperatures. These factors contribute to the extremely low water content in soil and reduce the survival of most of the microbial isolates with a role in the bioremediation of soils. We have developed a new technology based on the ability of anhydrobiotic microorganisms to withstand desiccation and the remarkable stability they display in a dried state. Because of their ability to survive without water and to promote plant growth under water stressing conditions, we have used this tolerance for the isolation of a collection of new desiccation-tolerant microorganisms that could be useful for the treatment of PAH-polluted soils in arid and semiarid regions.

Keywords: anhydrobiosis, polycyclic aromatic hydrocarbons, arid and semiarid soils, bioremediation, PGPR.

1 Introduction

Rhizoremediation combines the use of plants and microorganisms for the removal of pollutants. In this process, the microbes present in the rhizosphere of plants used during the bioremediation process provide an important contribution to the degradation of such pollutants [1, 2]. This technique has been extensively used for the treatment of soils polluted with polycyclic aromatic hydrocarbons

WIT Transactions on Ecology and the Environment, Vol 132, © 2010 WIT Press
www.witpress.com, ISSN 1743-3541 (on-line)
doi:10.2495/ETOX100121

(PAHs). This type of phytoremediation results from the higher densities and greater activities of microorganisms close to the plant's roots than in the surrounding soil [3, 4]. The combination of plants and microorganisms enhances the efficiency of contaminant removal thanks to the effect of root growth and its subsequent penetration through the different substrates and layers of soil. This allows the removal of entrapped contaminants that might have previously been inaccessible [5].

PAHs are recalcitrant pollutants often found in high residual concentrations in soils of industrial sites. This is a group of compounds of great environmental concern because some are mutagenic and carcinogenic [6]. PAHs are mainly introduced into the environment by emissions from human activities including power plants, industrial boilers, incinerators, ships, aircraft, automobiles and commercial and industrial heating systems [6]. Natural combustion such as forest fires and volcanic eruptions is another significant source of PAHs in soils [7, 8] Therefore, the soils of arid regions such as the Arabic peninsula or semiarid regions such as the Mediterranean basin are especially affected by PAH contamination and a lack of water [9].

Stressful conditions in arid and semiarid soils reduce microbial community diversity, mainly by limiting the horizontal migration of bacteria from the rhizosphere environment to the interspace because of the low moisture content and rapid water drainage in these types of soils [10].

There have been several studies of the most suitable plant species for the rhizoremediation of PAHs [11–13]. However, the presence and survival of beneficial rhizobacterial strains is very limited. The success of these beneficial processes is based on the rhizosphere competence of the microbes [14, 15], which is reflected by the ability of the microbes to survive in the rhizosphere, compete for the exudate nutrients, sustain in sufficient numbers and efficiently colonise the growing root system [14]. In our research group, we have developed a method for coating seeds with bacteria, a technique often used to apply beneficial microbes in a bioinoculant [16–19]. This approach is most successful when the bioinoculant is well adapted to the rhizosphere [20]; therefore, the isolation of well-adapted microorganisms to dry environments is of paramount importance for the successful rhizoremediation of arid and semiarid soils. In this work, we have developed a method for the isolation of desiccation-tolerant rhizobacteria and have shown their benefits as plant growth promoting rhizobacteria (PGPR) and their effect and survival under drought stress conditions.

2 Materials and methods

2.1 Soil samples

Soil samples were taken from the *Nerium oleander* rhizosphere subjected to seasonal drought at Granada (Spain) (37.182 N, 3.624 W) after a period of three months with no registered rainfall or any type of exposition to water. The soil sample was collected in a plastic bag, air-dried at room temperature, homogenised and sieved (2 mm mesh).

2.2 Microorganism, medium and culture conditions

Bacteria were grown in M9 minimal medium with PAHs as the sole carbon source at 30°C [21]. When PAHs were used, chips of naphthalene, anthracene, phenanthrene or pyrene were placed on the lid to avoid direct contact with cells; therefore, they were provided in vapour form.

2.3 Isolation of bacteria using a standard isolation method

One gram of air-dried soil was mixed with 10 ml sterilised water and thoroughly mixed. After the soil particles settled, serial dilutions were made and a 100 µl aliquot from each dilution were plated on M9 minimal medium plates. After 48 hours of incubation at 30°C, individual colonies were randomly picked and streaked out to obtain pure cultures. After the incubation time, the mixtures were transferred onto sterile glass plates and incubated in sterile conditions for 30 minutes to allow for the complete evaporation of water. The soil was mixed with water and diluted as described above. These tests were performed in triplicate. Alternatively, minimal medium with naphthalene, anthracene, phenanthrene or pyrene as the sole carbon source was used for the isolation of strains with biodegradative properties.

2.4 Sequencing of 16S rRNA genes and phylogenetic analysis

All strains isolated in this study were identified by the analysis of the partial sequence of the gene encoding 16S rRNA. Primers fD1, fD2, rD1 and rD2 [22] were synthesised by Sigma Genosis (UK) and used to amplify almost the full length of the 16S rRNA gene. Total DNA was isolated following the Kado and Liu method [23]. The PCR products were purified and sequenced as described previously by Pozo et al. [24]. The sequences were edited using 4Peaks software (http://mekentosj.com/4peaks/). Closely matched sequences were found in the GenBank database using the BLASTn algorithm [25].

2.5 Air drying: determination of survival rates

To determine survival rates, a colony of each pure culture containing 10^7–10^9 cells was resuspended in 1 ml of M9 minimal medium. Aliquot volumes (100 µl) were placed on sterile Petri dishes and dried under a current of sterile air for 24 hours. Cells were resuspended in 1 ml of sterile water, and serial dilutions of the cell prior and after drying were plated on TSA plates. All manipulations were performed at ambient temperature. The survival rate (%) was calculated as the rate of cells/ml after drying with reference to cells/ml before drying. The assays were performed in triplicate.

2.6 Sporulation test

To determine if the isolates were sporulant strains, a four-day-old colony was resuspended in 1 ml of sterile deionised water and incubated at 72°C for 30 min as described previously [17]. Aliquot volumes (100 µl) were plated before and after incubation. Strains able to grow in both conditions were discarded as sporulant or thermotolerant. Therefore, only temperature-sensitive strains were selected.

2.7 Plant growth promoting tests and plant protection against desiccation

Plants material and growth conditions consisted of a variation of the protocol established by Mayak et al. [26]. In this way, pepper (*Capsicum annuum* L. *cv.* Maor) seedlings were started from seeds that were sown in plastic trays in wet vermiculite. After one week, uniform-sized seedlings (shoot height of approximately 3 cm) were selected and planted in vermiculite, one per 7 cm diameter plastic pot. During the second week, the seedlings were fertilised once with 40 ml of either 1/10 or 1/5 Murashige and Skoog (MS) medium [27] as indicated. Three days after fertilisation, some of the seedlings were treated with 40 ml of bacterial suspension (A_{600nm} = 1.0), whereas others were watered with deionised water. Two weeks later, the seedlings were transplanted and watered. The seedlings were maintained in a growth chamber at a day/night temperature of 25/20°C with 25 µmol photons m^{-2} s^{-1} or 75 mol photons m^{-2} s^{-1} of light supplied for 12 h during the daytime.

For fresh (FW) and dry (DW) weight measurement, pepper plants were measured at 7, 14, 20 and 33 days after watering ceased. The relative water content (RWC) in pepper plants was also determined at the same times. The fully turgid weight (FTW), defined as the weight of the shoot after the plant had been held in 100% humidity conditions in the dark at 4°C for 48 h, of each plant was recorded. The RWC was calculated as follows:

$$RWC = \frac{FW - DW}{FTW - DW}$$

2.8 Statistical analysis

Data were analysed by analysis of variance (ANOVA), and pairwise comparisons were done using a student's *t* test. All hypotheses were tested at the 95% confidence level.

3 Results and discussion

The main objective of this study was the isolation of a collection of desiccation-tolerant microorganisms to study their effect on plants subjected to drought stress and their potential use for the removal of PAHs in rhizoremediation treatments.

3.1 Isolation of desiccation-tolerant microorganisms

For the isolation of desiccation-tolerant microorganisms a novel technique was developed by Narvaez-Reinaldo et al. [28] based on the differential survival of anhydrobionts in the desiccated state [17, 18]. Three different field samples of dry soil were collected from non-irrigated soils after three months without rain in Granada (South of Spain). In this area, there are two dry seasons, one during winter and another much drier season during summer. It was assumed that some of the environmental samples collected would be a good source of natural desiccation-tolerant microorganisms. Collections were made depending on the soil origin, including bulk soil, rhizoplane and rhizosphere. Soil samples were extensively dried as described in Materials and Methods. After the complete evaporation of water and suspension of the soil sample, serial dilutions were plated onto PAHs minimal medium plates. More than 36 strains from each plate were tested for their tolerance to withstand high temperature (as described in Materials and Methods), rendering 420 strains. Quick sporulation tests were performed for the selection of non-sporulating strains.

Because our main objective was to select desiccation-tolerant microorganisms, a series of tests to determine desiccation tolerance were performed to select 13 strains with a remarkable tolerance to drought and a survival of 4% or above to complete desiccation for 48 hours. Figure 1 shows the results of the desiccation tolerance tests.

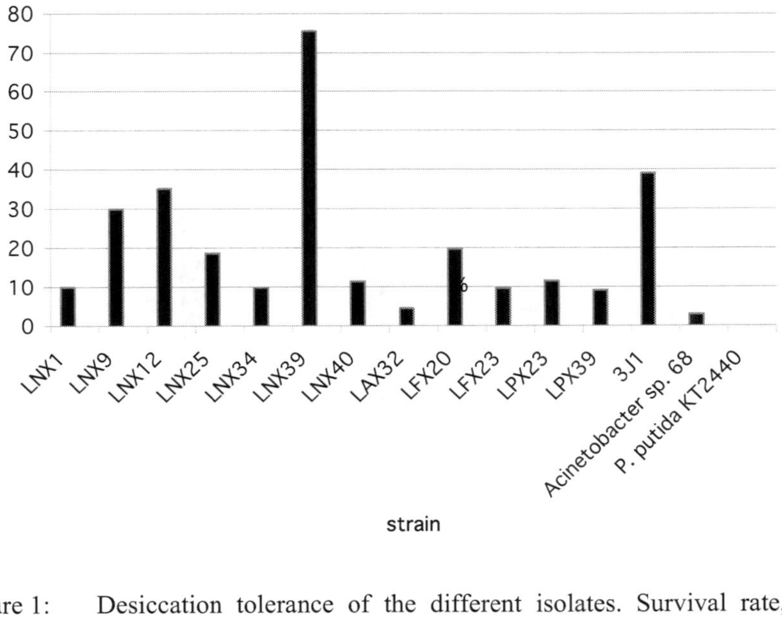

Figure 1: Desiccation tolerance of the different isolates. Survival rate, in percentage, is shown the y-axes. Name of the isolates is shown in the x-axes.

The molecular characterisation of the 14 different isolates based on the 16S rDNA homology of a partial sequence (1,440 bp) with the sequences in GenBank, the nucleotide sequencing of amplified 16S rDNA fragments obtained after colony PCR and the comparative analysis with the DNA databases allowed us to identify the desiccation-tolerant strains as members of the *Actinobacteria* group.

3.2 Plant growth promoting effect

A second desirable characteristic of this collection should be their ability to positively interact with a plant's roots. Some bacteria produce substances that stimulate the growth of plants through the production of phytohormones (i.e. auxins, gibberellins or cytokinins), ion uptake processes (such as iron that has been sequestered by bacterial siderophores or soluble phosphate), nitrogen fixation or the synthesis of plant development modulators such as the enzyme ACC deaminase, which can lower plant ethylene levels [29, 30].

We focused on selecting microorganisms promoting plant growth, in particular those protecting plants from desiccation. We tested the 13 desiccation-tolerant strains for their potential for growth promoting the three isolates that showed the highest level of desiccation tolerance: *Arthrobacter* sp., *Microbacterium* sp. and *Rhodococcus* sp. As a positive control, *Pseudomonas putida* KT2440, a saprophytic bacterium used as a model system for studying the biodegradation and interactions of a nonsymbiotic microorganism with plants,

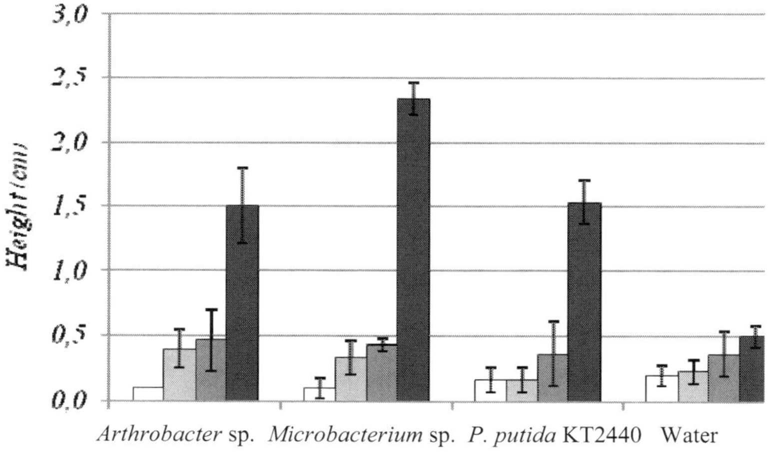

Figure 2: Stem size of pepper plants inoculated with the isolates *Arthrobacter* sp. and *Microbacterium* sp. Water was included as negative control and *P. putida* KT2440 as positive control. Measures were taken at 7 (white bars), 14 (light grey bars), 20 (dark grey bars) and 33 days (black bars).

was included [31, 32]. As a negative control, an equal volume of water was added instead. Figure 2 shows the effect of these three microorganisms on pepper plant growth. The addition of *Microbacterium* sp., *Arthrobacter* sp., and *P. putida* KT2440 produced an increase in plant growth determined by longer roots and stems and higher dry and fresh weights than negative controls. Plant growth was especially remarkable for plants inoculated with *Microbacterium* sp.

3.3 Plant protection against desiccation by bacterial strains

A similar test was performed with plants subjected to drought stress. Plants were inoculated with *Arthrobacter* sp., *Microbacterium* sp. and *Rhodococcus* sp. Additionally, plants inoculated with *P. putida* KT2440 and non-inoculated were also included. This time plant growth (roots and stem length and fresh and dry weight) was associated with time (7, 14, 20 and 33 days) in absence of water. Similarly, the FTW and RWC values were estimated. The results of these tests showed that only plants inoculated with *Arthrobacter* sp. and *Microbacterium* sp. presented a notable desiccation tolerance (Figure 3). These plants showed longer roots and aerial part of the plant, a higher fresh and dry weight and higher RWC, which is a value established by Mayak and co-workers to determine the tolerance of a plant to desiccation [26]. Therefore, these two strains were selected for future tests of rhizoremediation in arid and semiarid regions.

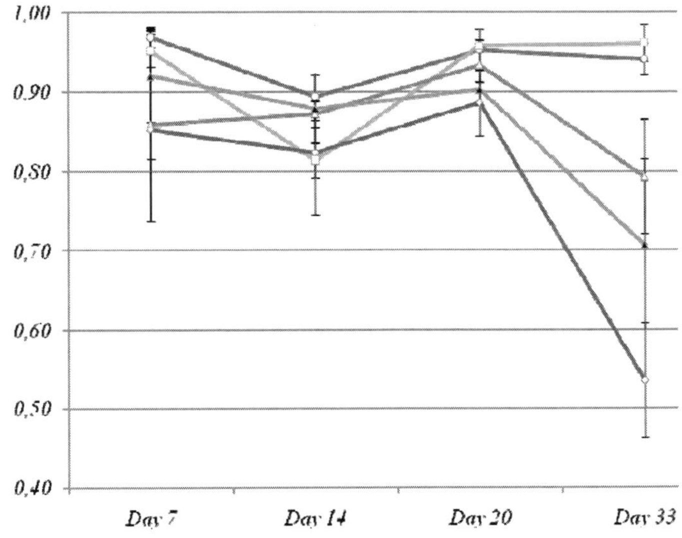

Figure 3: RWC values of pepper plants at 7, 14, 20 and 33 days of the drought stress inoculated with *Arthrobacter* sp. (circles), *Microbacterium* sp. (squares), *Rhodococcus* sp. (open triangles), *P. putida* KT2440 (diamonds), and water (solid triangles).

3.4 PAH removal by desiccation tolerant microorganisms.

Different isolates of *Arthrobacter* and *Microbacterium* have previously being used for the bioremediation of PAH-polluted soils [33, 34]. Therefore, we decided to test the ability of xerotolerant strains to grow on minimal media with naphthalene, anthracene, phenanthrene and pyrene as the only carbon source. The results of this test showed that the selected strains could grow on these PAHs. Therefore, the isolation of new strains was performed following the above mentioned method by plating in minimal medium with PAHs as the only carbon source. Using this method, 26 different strains were isolated in minimal medium with naphthalene as the sole carbon source, 10 strains were isolated using anthracene as the sole carbon source, 14 strains were isolated using phenanthrene as the sole carbon source and 13 strains were isolated using pyrene as the sole carbon source. We are currently assessing the efficiency of PAHs removal in liquid and in microcosms tests of those desiccation tolerance strains, as well as studying their ability to colonise roots. Future tests to establish the efficiency of the PAH removal of the bacterial strains and plant–microorganisms with the high and low presence of water are needed to determine which pair plant–microorganisms are the most appropriate for the efficient treatment of PAH-polluted arid and semiarid soils.

Acknowledgements

We thank the Junta de Andalucia (Spain) for funding this study through project reference P07-RNM-02588. Maximino Manzanera was granted by Programa Ramón y Cajal (Ministerio de Educación y Ciencia MEC, Spain and ERDF, European Union).

References

[1] Anderson, T.A., Guthrie, E.A. & Walton, B.T., Bioremediation in the rhizosphere. *Environmental Science & Technology*, **27** pp. 2630-2636, 1993.

[2] Schwab, A.P. & Banks, M.K., Biologically mediated dissipation of polyaromatic hydrocarbons in the root-zone. *Bioremediation through Rhizosphere Technology*, **563**, pp. 132-141, 1994.

[3] Cunningham, C.J., Ivshina, I.B., Lozinsky, V.I., Kuyukina, M.S. & Philp, J.C., Bioremediation of diesel-contaminated soil by microorganisms immobilised in polyvinyl alcohol. *International Biodeterioration & Biodegradation*, **54**, pp. 167-174, 2004.

[4] Ferro, A.M., Sims, R.C. & Bugbee, B., Hycrest crested wheatgrass accelerates the degradation of pentachlorophenol in soil. *Journal of Environmental Quality*, **23**, pp. 272-279, 1994.

[5] Parrish, Z.D., Banks, M.K., & Schwab, A.P., Assessment of contaminant lability during phytoremediation of polycyclic aromatic hydrocarbon impacted soil. *Environmental Pollution*, **137**, pp. 187-197, 2005.

[6] Baek, S.O., Field, R.A., Goldstone, M.E., Kirk, P.W., *et al.*, A review of atmospheric polycyclic aromatic-hydrocarbons - sources, fate and behavior. *Water Air and Soil Pollution*, **60**, pp. 279-300, 1991.

[7] Simoneit, B.R.T., A review of biomarker compounds as source indicators and tracers for air pollution. *Environmental Science and Pollution Research*, **6**, pp. 159-169, 1999.

[8] Nikolaou, K., Masclet, P. & Mouvier, G., Sources and chemical-reactivity of polynuclear aromatic-hydrocarbons in the atmosphere - a critical-review. *Science of the Total Environment*, **32**, pp. 103-132, 1984.

[9] Olivella, M.A., Ribalta, T.G., de Febrer, A.R., Mollet, J.M., & de las Heras, F.X.C., Distribution of Polycyclic aromatic hydrocarbons in riverine waters after Mediterranean forest fires. *Science of the Total Environment*, **355**, pp. 156-166, 2006.

[10] Dunbar, J., Takala, S., Barns, S.M., Davis, J.A., & Kuske, C.R., Levels of bacterial community diversity in four arid soils compared by cultivation and 16S rRNA gene cloning. *Applied and Environmental Microbiology*, **65**, pp. 1662-1669, 1999.

[11] Kuiper, I., Bloemberg, G.V. & Lugtenberg, B.J.J., Selection of a plant-bacterium pair as a novel tool for rhizostimulation of polycyclic aromatic hydrocarbon-degrading bacteria. *Molecular Plant-Microbe Interactions,* **14**, pp. 1197-1205, 2001.

[12] Qiu, X., Shah, S.I., Kendall, E.W., Sorensen, D.L., *et al.*, Grass-enhanced bioremediation for clay soils contaminated with polynuclear aromatic-hydrocarbons. *Bioremediation through Rhizosphere Technology*, **563**, pp. 142-157, 1994.

[13] Shann, J.R. & Boyle, J.J., Influence of plant-species on *in-situ* rhizosphere degradation. *Bioremediation through Rhizosphere Technology*, **563**, pp. 70-81, 1994.

[14] Lugtenberg, B.J.J. & Dekkers, L.C., What makes *Pseudomonas* bacteria rhizosphere competent? *Environ. Microbiol.*, **1**, pp. 9-13, 1999.

[15] Weller, D.M., & Thomashow, L.S., Current challenges in introducing beneficial microorganisms into the rhizosphere. *Molecular Ecology of Rhizosphere Microorganisms*, pp. 1-18, 1994.

[16] Tunnacliffe, A., Manzanera, M., Vilchez, S., & Garcia De Castro, A., Univ Cambridge Tech Services Ltd 2005.

[17] Vilchez, S., Tunnacliffe, A. & Manzanera, M., Tolerance of plastic-encapsulated *Pseudomonas putida* KT2440 to chemical stress. *Extremophiles*, **12**, pp. 297-299, 2008.

[18] Manzanera, M., Vilchez, S. & Tunnacliffe, A., Plastic encapsulation of stabilized *Escherichia coli* and *Pseudomonas putida. Applied and Environmental Microbiology*, **70**, pp. 3143-3145, 2004.

[19] Schippers, B., Scheffer, R.J., Lugtenberg, B.J.J. & Weisbeek, P.J., Biocoating of seeds with plant growth-promoting rhizobacteria to improve plant establishment. *Outlook on Agriculture*, **24**, pp. 179-185, 1995.

[20] Kuiper, I., Lagendijk, E.L., Bloemberg, G.V. & Lugtenberg, B.J.J., Rhizoremediation: A beneficial plant-microbe interaction. *Molecular Plant-Microbe Interactions*, **17**, 6-15, 2004.
[21] de Castro, A.G., Bredholt, H., Strom, A.R. & Tunnacliffe, A., Anhydrobiotic engineering of gram-negative bacteria. *Applied and Environmental Microbiology*, **66**, pp. 4142-4144, 2000.
[22] Weisburg, W.G., Barns, S.M., Pelletier, D.A. & Lane, D.J., 16S ribosomal dna amplification for phylogenetic study. *Journal of Bacteriology*, **173**, pp. 697-703, 1991.
[23] Kado, C.I., Liu, S.T., Rapid procedure for detection and isolation of large and small plasmids. *Journal of Bacteriology*, **145**, pp. 1365-1373, 1981.
[24] Pozo, C., Rodelas, B., de la Escalera, S. & Gonzalez-Lopez, J., D,L-Hydantoinase activity of an *Ochrobactrum anthropi* strain. *Journal of Applied Microbiology*, **92**, pp. 1028-1034, 2002.
[25] Altschul, S.F., Madden, T. L., Schaffer, A. A., Zhang, J. H., *et al.*, Gapped BLAST and PSI-BLAST: a new generation of protein database search programs. *Nucleic Acids Research,* **25,** pp. 3389-3402, 1997.
[26] Mayak, S., Tirosh, T. & Glick, B.R., Plant growth-promoting bacteria that confer resistance to water stress in tomatoes and peppers. *Plant Science*, **166,** pp. 525-530, 2004.
[27] Murashige, T. & Skoog, F., A revised medium for rapid growth and bio assays with tobacco tissue cultures. *Physiologia Plantarum*, **15**, pp. 473-&, 1962.
[28] Narvaez-Reinaldo, J.J., Barba, I., Tunnacliffe, A., Gonzalez-Lopez, J. & Manzanera, M., Rapid treatment method for isolation of desiccation tolerant strains and xeroprotectors. *Submitted.*
[29] Glick, B.R., The enhancement of plant-growth by free-living bacteria. *Canadian Journal of Microbiology,* **41**, pp. 109-117, 1995.
[30] van Loon, L.C., Plant responses to plant growth-promoting rhizobacteria. *European Journal of Plant Pathology*, **119**, pp. 243-254, 2007.
[31] Venturi, V., Zennaro, F., Degrassi, G., Okeke, B.C. & Bruschi, C.V., Genetics of ferulic acid bioconversion to protocatechuic acid in plant-growth-promoting *Pseudomonas putida* WCS358. *Microbiology-(UK)*, **144**, pp. 965-973, 1998.
[32] dos Santos, V., Heim, S., Moore, E.R.B., Stratz, M. & Timmis, K.N., Insights into the genomic basis of niche specificity of *Pseudomonas putida* KT2440. *Environ. Microbiol.*, **6**, pp. 1264-1286, 2004.
[33] Seo, J.-S., Keum, Y.-S., Kyu Cho, I. & Li, Q.X., Degradation of dibenzothiophene and carbazole by *Arthrobacter* sp. P1-1. *International Biodeterioration & Biodegradation*, **58,** pp. 36-43, 2006.
[34] Keuth, S., Rehm, H.J., Biodegradation of phenanthrene by *Arthrobacter polychromogenes* isolated from a contaminated soil. *Applied Microbiology and Biotechnology*, **34**, pp. 804-808, 1991.

An evaluation of organopollutant biodegradation by some selected white rot fungi: an overview

M. Tekere[1], J. S. Read[2] & B. Mattiasson[3]
[1]*Department of Environmental Sciences, School of Agriculture and Environmental Sciences, University of South Africa, South Africa*
[2]*School of Medicine University of Botswana, Gaborone, Botswana*
[3]*Department of Biotechnology, Lund University, Sweden*

Abstract

An evaluation of indigenous subtropical white rot fungi producing the ligninolytic enzymes laccase and manganese peroxidase only showed that the fungi could degrade synthetic dyes, PAHs and organochlorine pesticides. Biodegradation of the polymeric dye Poly 478 saved as a good screening indicator for organopollutant degradation by the fungi and based on Poly R478, the indigenous fungi; *Trametes versicolor, Trametes cingulata, Trametes pocas* and unidentified strain *DSPM95* were selected as high degraders from 11 screened fungi. PAH biodegradation was evaluated both in batch and continuous culture bioreactors. The degradation of PAHs at an initial concentration of 20 ppm each in static batch cultures over 31 days showed that + 60% fluorene, +40% phenanthrene, +42% anthracene and 3–11% of benzo(a)anthracene and pyrene were degraded. Biodegradation experiments using extra-cellular culture fluid demonstrated that the extracellular enzyme system of the fungi was responsible also for the biodegradation of the PAHs. Studies on PAH biodegradation by the isolate *DSPM95* in packed and suspended carrier continuous bioreactors also showed that the fungi could degrade most of the PAH fed continuously at a concentration of 1 ppm over a period of at least 31 days.

Keywords: white rot fungi, ligninolytic enzymes, polyaromatic hydrocarbons, synthetic dyes, organochlorine, biodegradation, bioremediation, manganese peroxidase, laccase.

WIT Transactions on Ecology and the Environment, Vol 132, © 2010 WIT Press
www.witpress.com, ISSN 1743-3541 (on-line)
doi:10.2495/ETOX100131

1 Introduction

Polyaromatic hydrocarbons and persistent organochlorine pesticides are some of the major groups of pollutants of concern due to their health effects, such as carcinogenicity, mutagenicity and reproductive effects (Asgher *et al.* [1]). Microbial biodegradation of organic chemicals is seen as an alternative to physical and chemical decontamination technologies. A large diversity of microorganisms exist in the environment where the pollutants exist and mostly upon exposure to the pollutants the microorganisms can use their metabolic machinery or evolve new metabolic mechanisms to breakdown the pollutants (Peng *et al.* [2]). Ligninolytic white rot fungi have been shown to degrade a wide range of pollutants including PAHs, synthetic dyes, pesticides and biphenyls (Asgher *et al.* [1]). Ligninolytic fungi use a system of peroxidases and oxidases to breakdown environmental pollutants. Liginin peroxidase, manganese peroxidase and laccases are the major ligninolytic enzymes and a number of accessory enzymes, such as H_2O_2-forming glyoxal oxidase, aryl alcohol oxidase, oxalate producing oxalate decarboxylase (ODC), NAD-dependent formate dehydrogenase (FDH) and P450 monooxygenase (Asgher *et al.* [1]), have been shown to play important roles in white rot fungi biodegradations. The advantages of a bioremediation system using white rot fungi is that they produce an extracellular, non-specific enzyme system for the degradation of diverse organopollutants, and they require no pre-conditioning to the pollutant for expression of the enzyme system and biodegradation. Although a lot of research progresss has been made in studying fungal applications in bioremediation, a lot still needs to be done in order to identify and quantify all active players from different ecosystems and their biopotentials. This paper gives an overview of the organopollutant biodegradation studies carried out on some white rot fungi from a subtropical forest in Zimbabwe.

2 Materials and methods

2.1 Sampling program

The fungal strain basidiocarps were collected in August 1995 from Chirinda, a moist evergreen forest, and Chimanimani (miombo woodland) forests in the Eastern Highlands of Zimbabwe. The fungi were growing on dead wood and were the dominant species at the time of collection and are well represented in Zimbabwean woodlands [3]. Pure cultures were obtained and maintained as described in Tekere *et al.* [4]. The pure isolates obtained were: *Trametes elegans (Berk.) Ryvarden, Trametes versicolor (Fr.) Pilat, Trametes cingulata (Berk., Trametes pocas (Berk.) Ryvarden, Isolate DSPM95*, Datronia concentric, Irpex sp*, Lentinus velutinus (Fr), Creptidotus mollis (Schaeff.:Fr.) Kummer* and *Pycnoporus sanguineus (Fr) Murill.* With the exception of *Irpex sp** and *DSPM95**, all the other isolates used were identified using gross and microscopic characteristics (Mswaka [3]).

2.2 Laboratory methods

2.2.1 Screening for organopollutant biodegradative activity

Lignin and synthetic dye biodegradations were used as determinants for pollutant biodegradation. Lignin biodegradations were carried out as in the method by Sundman and Nase [5] and described by Tekere *et al.* [4]. The decolourisation of the synthetic dyes was also determined on agar plates and blue dextran, crystal violet, cresol red and bromophenol blue were used (Sigma Chemical Co., St Louis, MO). The biodegradation experiments were carried out as described in Tekere *et al.* [4]. The decolourisation of Poly R478 dye was determined in liquid shake cultures using the same medium as described above with the agar excluded. Shake flasks (500ml), containing 100 ml of medium, were inoculated with four, 6 mm agar plugs and the flasks were incubated at 30°C on a shaker at 100 rpm. The decolourisation was determined over a 9 day incubation period by measuring the absorbance ratio 520/350 nm using a Shimadzu Biospec-1601 (Tekere *et al.* [4]).

Shallow stationery batch cultures were used for the production of the ligninolytic enzymes, lignin peroxidase, manganese peroxidase and laccase using a base medium as proposed by Bonnarme *et al.* [6]. Culturing and enzyme assays were carried out as described in Tekere *et al.* [7].

2.2.2 Polycyclic aromatic hydrocarbon biodegradation

The biodegradation of representative 2-, 3- and 4- ringed polycyclic aromatic hydrocarbons (PAHs) fluorene, phenanthrene, anthracene, pyrene and benzo(a)anthracene was studied in static batch cultures of the isolates; *P. chrysosporium, T. versicolor, T. cingulata, T. pocas* and *DSPM95*. Apart from *P. chrysosporium*, the fungal isolates were selected from the indigenous Zimbabwean isolates on the basis of high ligninolytic and high dye decolourisation activities from preliminary screening. Mineral medium as in Tekere *et al.* [8], was used for culturing the fungi and a mixture of the PAHs at a concentration of 20 mgl^{-1} for each of the PAHs in a 1 ml acetone stock solution was used in 50 ml of the medium in 500 ml shake flasks. The degradation experiments were carried out as described in Tekere *et al.* [8]. The influence of the addition of glucose oxidase and additional glucose on the biodegradation of the more recalcitrant pyrene and benzo(a)anthracene and the lower molecular weight anthracene was also studied. Glucose oxidase, type V-S from *Aspergillus niger,* (Sigma Chemical Co) and additional glucose, 5gl^{-1} were added to cultures after 14 days of incubation. Sampling and PAH extractions and analysis were done as described in Tekere *et al.* [8]. Effects of addition of glucose oxidase and additional glucose were determined as the hydrogen peroxide - dependant oxidation of phenol red (Tekere *et al.* [8]).

2.2.3 PAH biodegradation by fungal extracellular supernatants

Culture supernatants from 3 week old batch cultures were aseptically separated from the mycelia, centrifuged at 6500 rpm, filtered through 0,2 μm sterile syringe filters and used for the biodegradation experiments (Tekere *et al.* [8]). The biodegradations were carried out in triplicate for 6 days for each isolate, sampling time and analysis and different experimental conditions and 10 mgl^{-1} of each of the PAHs were used. Samples were frozen at intervals and used for residual PAH analysis. Residual MnP and laccase activity was also determined.

2.2.4 PAH biodegradation in packed bed and suspended carrier bioreactors

The white rot fungal isolate *DSPM95* was selected for PAH biodegradation studies in packed bed and suspended carrier bioreactor systems on the basis of high dye degradation activity, high ligninolytic activity and high PAH biodegradation activity in batch cultures. A jacketed glass column with a working volume 500 ml, filled up with Poraver glass bead carriers which are finely pored glass granules (8 - 16 mm pore size), made from purified recycled glass (Dennert Poraver GMBH, Germany) was used for the packed bed reactor. A jacketed glass column with a working volume of 1000 ml, filled with 550 ml kalderns plastic carriers (Anoxa, Lund, Sweden) was used as a suspended carrier bioreactor system. The plastic carriers are cylindrical, with an inner diameter of 8 mm, which is sub-divided into four, and have a rough outside wall allowing for good cell attachment. The assembled reactors were sterilised, inoculated and operated as described in Tekere *et al.* [9]. Residual PAHs, MnP and laccase activities in the system were followed from the reactor outflow.

2.2.5 Polycyclic aromatic hydrocarbon extraction and analysis

Residual PAHs in the samples from both the bioreactors and batch cultures were extracted with hexane and the extraction and analysis of the PAHs is described in Tekere *et al.* [8] and Tekere *et al.* [9]. External standards were used for the quantitative determination of the residual PAHs and the detection limit achieved was 0.01 mgl^{-1} for all the compounds. The PAH recovery efficiency for the extraction method was 97 – 99% for fluorene, 96% for phenanthrene, 99% for anthracene, 94% for pyrene and 91% for benzo(a)anthracene. The results are represented as means of the triplicate sample extractions and analyses. The standard deviations were calculated wherever applicable. PAH metabolites were partially identified using GC-MS (Varian Star, 3400, Varian Saturn II) as described in Tekere *et al.* [8]. A PAH search library (supplied by Varian chromatography systems, Sweden) was used for the qualitative metabolite identification.

2.2.6 Biodegradation of organochlorines by the fungi

The organochlorines, lindane, dieldrin, dicofol and endosulfan at a concentration of 20 ppm were incorporated as a mixture into 50 ml growth medium in 500 ml Erlenmeyer flasks. The organochlorines used in these experiments were of technical grade and active ingredient analysis by gas chromatography gave 93%, 95%, 49% and 33% for dicofol, lindane, endosulfan and dieldrin respectively. As controls for non-biological loss, controls in which the pesticides were added to uninoculated contaminated medium were used and both control and test experiments were carried out in the same way as for the fungal PAH degradations described above.

Samples were analysed for residual pesticides using gas chromatography. Sample (0.5 ml) was macerated with 5 ml dichloromethane in a macerating cup for 5 min after which the sample was filtered through sodium sulphate. The dichloromethane was evaporated and the samples were reconstituted in 2 ml hexane for gas chromatography analysis. GC (Varian Star 3400 Cx), fitted with an electron capture detector (ECD) and a capillary column QC 5/BP 10, 25 cm was used. The detector temperature was set at 300°C, injector temperature 200°C and column temperature was 195°C. External standards were used in the analyses.

3 Results and discussions

As shown in table 1 and fig 1, all isolates had some lignin and dye decolourisation activity demonstrating potential for pollutant degradation. The isolates *DSPM95* and *T. versicolor* had the highest degradative abilities. No lignin peroxidase activity was detectable in any of the screened isolates used except in the reference strain, *P. chrysosporium*. Even when Mn^{2+}, which is known to reduce LiP titers was excluded (Bonnarme *et al.* [6]) or when veratryl alcohol was added as an inducer, no LiP could be detected. Laccase and manganese peroxidase activities were detectable in all the tested fungal species and the activity levels were variable within the isolates, with *C. mollis* showing little detectable laccase and manganese peroxidase activities while *T. versicolor* and *DSPM95* showed the highest enzyme activities (Tekere *et al.* [7]).

i) Total PAH degradation in static batch cultures, ND – not detectable,

ii) Total degradation of anthracene, pyrene and benzo(a)anthracene only in static batch cultures at 31 days. The standard deviation was calculated from 3 - 6 samples (Tekere *et al.* [8]).

Table 1: Decolourisation magnitudes for the dyes and ligninolysis by the fungal isolates. Extent of decolourisation, scale 0–5 with decolourisation maximum at 5, as assessed by the area and magnitude of decolourisation. Lignin degradation, (+), (++), (+++) – in order of increasing ligninolysis magnitude.

Isolate	Lignin	Dye degradation			
		C.violet	B.blue	C.red	B.dextran
DSPM95	+++	5	5	5	5
T. versicolor	+++	5	5	5	5
T. pocas	+	4	3	5	2
T. cingulata	+	2	5	4	5
T. elegans	+	1	3	3	3
D.concentrica	+++	2	5	3	3
Irpex spp	++	1	5	5	3
L.velutinus	+	3	4	2	1
C.mollis	+	3	1	3	1
P. sanguineus	+	2	3	3	2

Tekere *et al.* [4].

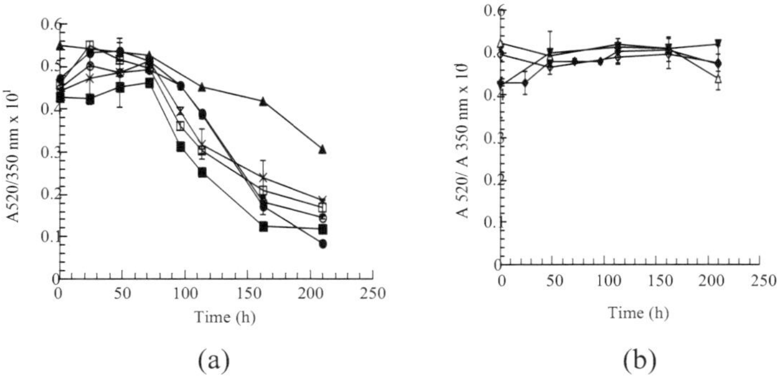

Figure 1: Poly R478 decolourisation by white rot fungal isolates in liquid shake cultures (100 rpm) as a function of time. Decolourisation was determined by the absorbance ratio A520/350 nm: —○—*T. pocas*,—●—*T. versicolor*, —□—*T. cingulata*, —▲— *D. concentrica, Isolate DSPM95* —×— *P. sanguineus*, —◇—*C. mollis*, —△— *L. velutinus*, —▼—*Irpex spp and* —◆—*T. elegans* (Tekere *et al.* [4]).

Table 2: Total PAH degraded by the fungal isolates (homogenised whole cultures analysis) as a percentage of the initial concentration. The cultures were incubated at 30°C for 31 days.

Isolate	Fluorene	Phenan-threne	Anthracene	Pyrene	Benzo(a anthracene
P.chryso-soporium	i.76.0±6.7	41.7 ±1.2	78.0± 10.0	15.0 ± 5.4	16.0±2.5
			ii.90.0 ± 1.6	13.0 ± 2.0	15.0±0.4
DSPM95 spp	i.82.0±3.1	57.0 ± 2.7	85.0 ± 10.6	11.0±0.5	12.0±2.8
			ii.98.0 ± 0.5	52.0±5.3	33.0±4.5
T. versicolor	i.78.0±5.5	49.0 ± 7.4	54.0 ± 7.0	ND	3.0 ± 0.0
			ii.82.0±13.0	9.0 ±6.8	30.0±1.4
T. pocas	i.72.0±6.3	60.0±22.0	70.0 ± 2.6	8.0 ±0.6	9.0 ± 5.4
			ii.99.0 ± 0.2	35.0±2.2	38.0±2.0
T. cingulata	i.63.4±1.6	42.0 ± 3.4	9.0 ± 2.1	9.0± 2.1	10.0±4.7
			ii.99.0 ± 0.6	40.0±10	23.0±5.6

The isolated fungi had high degradative activity for flourene, anthracene and phenanthrene and degradation was low for high molecular weight PAHs, pyrene and benzo(a) anthracene. PAH adsorption to fungal mycelium occurred and as shown in table 3, high pyrene and benzo(a)anthracene remained associated with mycelium at the end of the culturing period.

PAH biodegradation in fungal extracellular fluids under different experimental treatments showed that fungal secretions in the extracellular media was responsible for the biodegradation and addition of H_2O_2/ Mn^{2+} and/ ABTS caused an increase in the degradation of the PAHs in most cases though the effects were variable for different fungi. Anthracene was most degraded in

Table 3: The percentage PAH associated with the mycelia of the fungal strains. This percentage PAH sorption was calculated from the difference in the PAH degraded as analysed from the culture supernatant extractions and the PAH degraded as analysed from the homogenised whole culture extractions.

Isolate	% PAH associated with the mycelia				
	Fluorene	Phenanthrene	Anthracene	Pyrene	Benzo(a) anthracene
P. chrysosporium	10.0±2.0	7.0± 0.0	16.0± 5.0	54.0± 6.4	52.0±10.0
DSPM95	6.0 ± 0.8	22.0 ± 3.5	8.0 ± 3.4	40.0±11.0	53.0 ± 5.5
T. versicolor	5.0±0.15	25.0 ± 5.0	26.0 ± 8.8	36.0 ± 1 6	32.0 ± 23
T. pocas	10.0±6.0	33.0 ± 5.1	26.0 ± 0.5	54.0 ± 0.5	33.0± 11.0
T. cingulata	4.2 ± 2.0	22.0 ± 10.0	10.0 ± 3.0	36.0 ± 2.0	46.0 ± 8.4

Tekere *et al.* [8]

Table 4: PAH biodegradation in extracellular fluids under different experimental treatments, incubated at 30°C for 6 days.

Isolate	Culture treatment	% PAH degraded				
		Fluoren e	Phenanth rene	Anthra cene	Pyrene	Benzo(a) nthracen e
P.chryso-soporium	Culture fluid	28.0	15.0	17.5	8.0	4.9
	+ H_2O_2/ Mn^{2+}	69.0	32.5	66.5	36.0	22.0
	+H_2O2/Mn^{2+} /ABTS	41.0	37.5	16.0	29.0	23.0
DSPM95 spp	Culture fluid	56.0	29.5	55.0	15.0	25.0
	+ H_2O_2/ Mn^{2+}	72.0	35.0	79.0	22.0	53.0
	+H_2O2/Mn^{2+} /ABTS	71.0	43.0	63.0	37.5	32.0
T. versicolor	Culture fluid	42.0	19.5	23.5	7.0	8.5
	+ H_2O_2/ Mn^{2+}	77.5	37.5	43.0	20.0	52.5
	+H_2O_2/Mn^{2+}/ABTS	70.0	39.0	30.5	31.5	58.0
T. pocas	Culture fluid	38.5	23.0	29.5	17.0	18.2
	+ H_2O_2/ Mn^{2+}	64.0	38.0	77.0	19.5	30.5
	+H_2O_2/Mn^{2+}/ABTS	67.0	61.0	32.0	33.0	38.0
T.cingulata	Culture fluid	24.0	19.5	22.0	12.0	13.0
	+ H_2O_2/ Mn^{2+}	71.0	32.5	32.0	19.5	18.0
	+H_2O_2/Mn^{2+}/ABTS	66.5	35.0	22.5	15.5	21.0

conditions with H_2O_2/ Mn^{2+} while phenanthrene was most degraded under H_2O_2/ Mn^{2+} and/ ABTS conditions. In controls only 20.5% and 9.2% of fluorene and phenanthrene respectively was not recovered, thus attributed to abiotic loss, all of the other PAHs was recovered from the controls. Determination of residual enzyme activity in the extracellular fluid of the fungal cultures showed that the enzymes remained relatively stable over an 8-day incubation period at 92% for MnP and 88% for laccase.

Metabolic intermediates occurred during PAH biodegradation and several metabolic peaks were evident during PAH analysis. Comparison of the pattern of PAH degradation in *P. chrysosporium* and *T. pocas, T. cingulata, T.*

versicolor and *DSPM95* based on GC-MS pattern showed that there was more metabolite breakdown by the other fungi than *P. chrysosporium* at 31 days. As an example Fig 2a and b shows the GC-MS pattern for *T. pocas* and *P. chrysosporium* respectively at 31 day incubation time. This shows that different fungi should be explored in order to come up with the best candidates for complete degradation of the PAHs.

Evaluation of PAH biodegradation in both the packed bed and suspended carrier bioreactors showed that high degradation and enzyme activities could be achieved with immobilised fungi. Since fungi are slow growing microorganisms and prefer growing attached to surfaces, retaining them in immobilised bioreactors is an advantage. Fig 3a and a show the PAH biodegradation and enzyme activities for *DSPM95* in the packed bed bioreactor. A similar profile was obtained for the suspended carrier bioreactor as presented in Tekere *et al.* [9]. Both reactor configurations could be used in bioremediation of polluted waste effluent, the suspended carrier however offers the advantage of good fluid mixing and no clogging with long term operation like what happens with the void pore spaces in packed bed reactors.

Control reactor studies showed that PAH recovery was low in the initial stages of the reactor operation but PAH recovery at a fixed flow rate increased with time while PAH recovery at different flow rates, increased with increased flow rate. For example the PAH recovered increased with time from 30.0% to

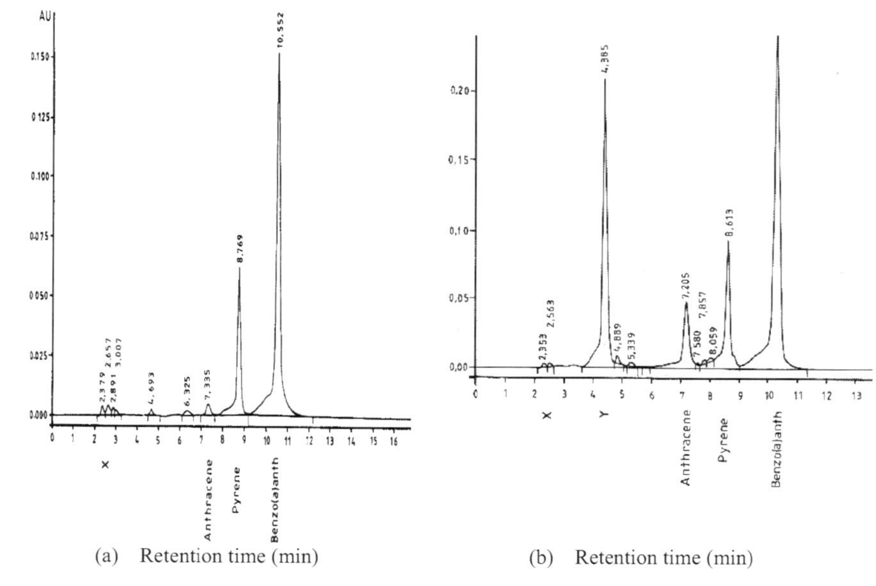

Figure 2: Chromatograms for the biodegradation of anthracene, pyrene and benzo(a) anthracene in 31 day old cultures of (a) *T. pocas and* (b) *P. chrysosporium* with glucose and glucose oxidase added. x and y are metabolic peaks (Tekere *et al.* [8]).

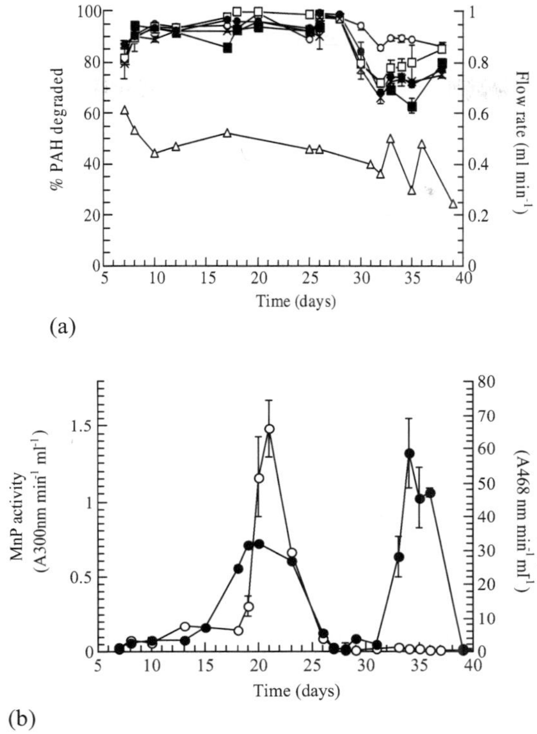

Figure 3: Biodegradation of the PAHs by *DSPM95* and changes in enzyme activities and medium flow rates in the packed bed bioreactor system. (a) —○— fluorene, —■— phenanthrene, —□— anthracene, —×—pyrene, —●— benzo(a) anthracene and —△— flow rate. (b) —●— MnP and —○— laccase activity. The reactor operation conditions were: (i) day 7–13 basal feed medium with no glucose was fed; (ii) day 13–29 medium with 2 g glucose, 25 ppm Mn^{2+} and 0.66 g N (a medium optimum for enzyme production for the isolate *DSPM95*); (iii) day 29–36 medium with no glucose, 25 ppm Mn^{2+} and 0.66 g N; (iv) day 36–40 medium with 2 g glucose, 25 ppm Mn^{2+} and 0.66 g N (Tekere *et al.* [9]).

almost 80.0% for the packed bed reactor at a flow rate of 0.5 ml min^{-1}. This could be attributed to the saturation of the absorption capacity of the carriers. Determination of the PAH adsorbed to the carrier material revealed that around 2.0% flourene, phenanthrene and anthracene, and 21.0% and 13.0% pyrene and benzo(a)anthracene respectively could be recovered from the carriers. Airlift of the PAHs from the control reactors also occurred. PAH lost from the medium in

Table 5: Percentage PAH degraded at different flow rates in the suspended carrier bioreactor.

Flow rate ($mlmin^{-1}$)	% PAH degraded				
	Fluorene	Phenathrene	Anthracene	Pyrene	Benzo(a) anthracene
0.5	94.7± 0.7	94.0 ± 1.0	97.0 ± 0.1	92.0 ± 2.0	91.0 ± 1.0
0.82	95.6± 2.0	93.1 ± 0.9	96.9 ± 0.9	92.4 ± 3.1	92. 0 ± 0.5
1.1	75.4± 6.0	83.9 ± 1.5	89.5 ± 3.2	85.0 ± 11	85.1 ± 3.0

Results are given as % mean recovery from at least 2 to 4 samples with standard deviation. Tekere *et al.* [9].

Table 6: Lindane, dicofol, endosulfan and dieldrin degradation in static batch cultures of some white rot fungi in 31 days.

Isolate	% Pesticide degraded			
	Lindane	Dicofol	Endosulfan	Dieldrin
DSPM95	84.0	95.0	97.0	98.0
T. pocas	99.0	97.0	98.0	98.0
P. sanguineus	99.0	99.0	99.0	98.0
D.concentrica	93.0	99.0	98.0	98.0
T. versicolor	61.0	ND	50.0	30.0
T. cingulata	62.0	ND	50.0	99.0

ND - pesticide degradation not well determined.

the suspended carrier bioreactors through airlift at a airflow rate of 1.4 l min^{-1} after 48 h was 43%, 36%, 32%, 27% and 26% for flourene, phenanthrene, anthracene, pyrene and benzo(a)anthracene respectively. However it should be noted that in inoculated reactors adsorbed PAHs will be degraded and the growing biomass can contribute to reduction of the PAHs lost though aeration.

The degradation of the organochlorines; lindane, dicofol, endosulfan and dieldrin by the fungal isolates is shown in Table 6. With the exception of the isolates *T. versicolor* and *T. cingulata* all the other fungal isolates consistently degraded the pesticides dicofol, endosulfan and dieldrin by above 95.0% of the initial parent compounds showing their potential in pesticide bioremediation. Further studies on lindane degradation are reported in Tekere *et al.* [10].

4 Conclusions

The ability of the white rot fungi to degrade a wide range of pollutants means that they can be best for application in the biodegradation of mixed pollutants in soil and wastewater, where a considerable diversity of organopollutants is often found. The white rot fungi reported here can be explored further for bioremediation of synthetic dyes, PAHs and organochlorine pesticides among other pollutants thus offering a solution in the clean-up of the environments contaminated by these pollutants. Continuous screening of mainly those basidiomycetes white rot fungi which have not yet been studied can yield isolates which can produce not only high ligninolytic enzyme levels but can also cause complete pollutant degradation. This study also shows that some of the studied white rot fungi with only MnP and laccase activities actually have higher biodegradation capabilities than the model fungus *P. chrysosporium* which is a LiP and MnP producer. While the model fungus *P. chrysosporium* showed the accumulation of anthraquinone, the degradative pathways for PAHs by selected sub - tropical white rot fungi showed further metabolism of the intermediate metabolites such as anthraquinone. The white rot fungi can also be used successfully in continuous packed bed and suspended carrier bioreactors to achieve high and prolonged pollutant degradation.

References

[1] Asgher, M., Bhatti, H.N., Ashraf, M. & Legge, R.L. Recent developments in biodegradation of industrial pollutants by white rot fungi and their enzyme system. *Biodegradation*, **19**, pp. 771-783, 2008.

[2] Peng, R., Xiong, A., Xue, Y., Fu, X., Gao, F., Zhao, W. & Tian, Y. Microbial degradation of polyaromatic hydrocarbons. *FEMS Microbiology Reviews* **32**, pp. 927-955, 2008.

[3] Mswaka, A.Y. Studies on Trametes occurring in indigenous forests of Zimbabwe. PhD Thesis, Cranfied University, UK, 1995.

[4] Tekere, M., Mswaka, A.Y., Zvauya, R. & Read, J.S. Growth, dye degradation and ligninolytic activity studies on Zimbabwean white rot fungi. *Enzyme and Microbial Technology*, **28**, pp. 420 – 426, 2001.

[5] Sundman, V. & Nase, L. A simple plate test for direct visualisation of biological lignin degradation. *Paper and Timber*, 2, pp. **67** – 71, 1971.

[6] Bonnarme, P., Perez, J. & Jeffries, T.W. Regulation of ligninase production in white rot fungi. In Leatham, G.F. and Himmel, M.E. (Eds.), *Enzyme in Biomass Conversion, ACS symposium Series 460*, ACS, Washington, D.C. pp. 200 – 206, 1991.

[7] Tekere, M., Zvauya, R. & Read, J.S. Ligninolytic enzyme production from selected sub - tropical white rot fungi under different culture conditions. *Journal of Basic Microbiology*, **41**, pp. 97 – 111, 2001.

[8] Tekere, M., Read, J.S. & Mattiasson, B. Polycyclic aromatic hydrocarbon biodegradation in extracellular fluids and static batch cultures of selected Sub - Tropical white rot fungi. *Journal of Biotechnology*, **115**, pp. 367-377, 2005.

[9] Tekere, M. Read, J.S. & Mattiasson, B. Polycyclic aromatic hydrocarbon Biodegradation by a sub - tropical white rot fungus in packed-bed and suspended carrier bioreactor systems. *Environmental Technology*, **28** pp 683 – 691, 2007.

[10] Tekere, M., Ncube, I., Read, J.S. & Zvauya, R. Biodegradation of the organochlorine pesticide, lindane by a tropical white rot fungus in batch and packed bed bioreactor systems. *Environmental Technology*, **23**, pp. 199–206, 2002.

Adaptation of bacterial biotests for monitoring mycotoxins

Cs. Krifaton[1], J. Kukolya[3], S. Szoboszlay[2], M. Cserháti[2], Á. Szűcs[1] & B. Kriszt[2]
[1]*Szent István University, Regional Center of Excellence, Hungary*
[2]*Szent István University,*
Environmental Protection and Environmental Safety, Hungary
[3]*Agruniver Holding Ltd., Hungary*

Abstract

Mycotoxins are secondary fungal metabolites that have mutagenic, carcinogenic, teratogenic, immunomodulant and cytotoxic effects. Besides expensive chemical analytical methods, biotests can be alternative screening methods for selecting mycotoxin degrading microbes. Our aim was to monitor mycotoxins by bacterial biotests. Aflatoxin-B1 (AFB1), deoxynivalenol, zearalenon, T2-toxin and ochratoxin were analysed by two biotests: *Aliivibrio fischeri* luminescence assay and SOS-Chromotest using *Escherichia coli*. The *A. fischeri* bioluminescence assay is one of the most sensitive bacterial toxicity assays across a wide spectrum of toxicants; however, the effects of mycotoxins have been hardly examined. The inhibition effect of various mycotoxins on *A. fischeri* was determined in the range of 1–20 μg/ml toxin concentration. The test bacterium proved to be the most sensitive for AFB1. Less but pronounced inhibition of zearalenon was also determined, while other mycotoxins had no measurable toxic effect on *A. fischeri*. Based on these results we have adopted this method for monitoring microbial AFB1 biodegradation efficiency. The genotoxic effect of mycotoxins was analysed by SOS-Chromotest. The principle of the assay is the SOS response that is induced by DNA-damaging agents. We have examined AFB1 0,078–10 μg/ml concentration. We found that the genotoxic effect of AFB1 is still detectable at 0,078 μg/ml concentration. SOS-Chromotest was successfully used to select microorganisms that have the best AFB1 degrading potential and can degrade AFB1 without genotoxic by-products. The applicability of SOS-Chromotest for screening mycotoxin degrader microbes

WIT Transactions on Ecology and the Environment, Vol 132, © 2010 WIT Press
www.witpress.com, ISSN 1743-3541 (on-line)
doi:10.2495/ETOX100141

was confirmed by parallel ELISA and chemical analytical tests. We have adopted two bacterial biotests that are suitable for monitoring mycotoxin concentrations based on their biological effects. On the basis of these results a screening method was developed for microbes with AFB1 degrading potential. Furthermore, these methods can be appropriate tools for elucidating mycotoxin metabolism by high throughput screening of expression, and transposon mutagenesis clone libraries of the best AFB1 degrading strains.

Keywords: aflatoxin-B1, zearalenon, ochratoxin, T2-toxin, deoxynivalenol, SOS-Chromotest, genotoxicity, Aliivibrio fischeri, screening, biodegradation.

1 Introduction

Mycotoxins are very complex micropollutants that can be found in every variety of grain and forage produced for food or feed. The most considered mycotoxins are aflatoxins (mainly B1) produced by *Aspergillus spp.*, zearalenon, deoxynivalenol, T2-toxin and ochratoxin produced by *Fusarium spp.* These substances are secondary fungal metabolites that have mutagenic, carcinogenic, and teratogenic, immunomodulant and cytotoxic effects [1], thus biomonitoring of mycotoxins has an increasing importance nowadays. Besides expensive chemical analytical methods, biotests can be alternative screening methods for selecting mycotoxin degrading microbes as they provide prompt information, moreover they are reliable and cost effective methods [2]. Our aim was to monitor mycotoxins – aflatoxin B1 (AFB1), zearalenon (ZEA), deoxynivalenol (DON), T2-toxin (T2) and ochratoxin (OCHRA) – by bacterial biotests. A bioluminescence assay based on *Aliivibrio fischeri* marine bacterium is a widely used bacterial biotest for ecotoxicological testing in environmental industry. The test bacterium emits light in optimal conditions; this property is used to determine ecotoxicity in environmental biotechnologies [3–6]. The *A. fischeri* bioluminescence assay is one of the most sensitive bacterial cytotoxicity assays across a wide spectrum of toxicants [7–9]. The specific *A. fischeri* strain, NRRLB-11177, has been widely used for acute toxicity estimation and several commercial test kits, i.e., Microtox, LUMIStox and ToxAlert are based on this strain [10]. However, the effects of mycotoxins have been hardly examined. SOS-Chromotest, a simple bacterial colorimetric assay, was used in this study for testing genotoxicity. Scientific literature has reported the sensitivity of this test for AFB1 [11–13] and high correlation (60–100%) with the widely used Ames test [12, 14–19]. Our aim was to develop bacterial biotests for monitoring mycotoxins and screening mycotoxin degrading microbes.

2 Methods

2.1 SOS-Chromotest

The test uses the PQ37 mutant strain of *Escherichia coli* wild type K12. The principle of the assay is that most genotoxins induce SOS response in the test bacteria. The test took advantage of an operon fusion placing lacZ, the structural gene for β-galactosidase, under the control of the sfiA gene, which is involved in cell division and make up – with another 17 genes – giving the SOS-error-prone. Thus β-galactosidase activity highly depends on sfiA expression [20]. We have carried out SOS-Chromotest according to Legault et al. [21]. SOS-Chromotest was originally designed as a test tube procedure to detect DNA damage [11]. This test makes it possible to use S9 rat liver homogenates to examine those compounds that have indirect genotoxic activity and need metabolic activation to actuate their genotoxic effect. The SOS-Chromotest Kit was purchased from Environmental Bio-Detection Products Inc., Canada. The test was conducted according to the manufacturer's instructions. We have tested AFB1, ZEA, OCHRA, T2 and DON. 10 μl of AFB1 (10 μg/ml) in two fold serial dilution and 10 μl of ZEA, DON, T2 and OCHRA in two concentrations (10 and 5 μg/ml) were analysed. Enzyme activity (β-galactosidase and alkaline-phosphatase relative concentrations) were measured by ELx800 (BioTek Instruments, Inc.) at 405 and 620 nm. Induction factors (IF) indicative of genotoxicity were calculated by the following formula, eqn (1):

$$I = (A_{405}\,nc \times A_{620}\,t)/(A_{405}\,t \times A_{620}\,nc) \qquad (1)$$

where nc is the negative control and t is the tested sample. An induction factor of 1,5 or more is determined as genotoxic [11].

2.2 *A. fischeri* luminescence assay

An acute bioluminescence assay was modified according to Sarter et al. [22]. The bioluminescent marine bacteria *Aliivibrio fischeri* (DSM-7151, NRRLB-11177) was purchased from DSMZ. Cultures of the microorganism were stored on Bactomarine slant agar (Difco, USA) at 4°C. Bactomarine broth (Difco, USA) was inoculated with 24 hour colonies from Bactomarine slant agar. The experiment was carried out in liquid cultures and luminescence was determined by a Toxalert 100™ (Merck, Germany) luminometer (Merck KGaA, Germany) after 3.5, 10, 15, 25 h of incubation (25°C, 30 rpm in shaking thermostat) according to Froehner et al. [23] by the following formula:

$$\text{Inhibition (\%)} = ((C_{tx}-S_{tx})100)/C_{tx} \qquad (2)$$

where C_{tx} gives the arithmetic mean of the bioluminescence values of parallel controls after the examined hour (x = contact time in point) and S_{tx} represents the bioluminescence average value of parallel samples determined at contact time in point.

2.2.1 Mycotoxin monitoring with *A. fischeri*

Basic experiments to investigate the sensitivity of the test-organism for mycotoxins, such as AFB1, ZEA, T2, OCHRA, and DON, were performed in 20 ml cultures of *A. fischeri*, in which optical density was adjusted to 0,1 at 550 nm (corresponding to 10^6 CFU/ml). Mycotoxins were purchased from Fermentec Ltd. (Israel) and were dissolved in acetone. Controls without mycotoxins contained the culture plus acetone. Tests with controls, AFB1 (20, 10, 5, 2, 1 μg/ml), ZEA, T2, OCHRA and DON (20, 15, 10 μg/ml) were conducted in triplicate.

2.3 Biodegradation tests

According to published results [24] actinomycete strains – especially *R. erythropolis* strain DSM14303 [25] – are the best aflatoxin degraders. Based on these findings ten rhodococcus strains belonging to four species (*Rhodococcus erythropolis, R. ruber*, *R. globerulus* and *R. rhodochrous*) were used for the biodegradation experiment. Rhodococci from the strain collection of Agruniver Holding Ltd. (*R. erythropolis* strains N11, OM72, ZMF231, *R ruber* N361, *R. globerulus* N58 and *R. rhodochrous* K402) and reference strains from international collections (*R. erythropolis* DSM 4306, NCIMB9784, IFO12538, *R. rhodochrous* CW25 and ATTC12674) were chosen. The *Escherichia coli* strain K12 was also used as a non mycotoxin degrader control microbe. Cultures were streaked on LB agar and were incubated at 28°C for 72 hours. After checking purity, cells were inoculated into 50 ml LB media and were incubated at 170 rpm, 28°C for 72 hours. Absorbance of stock cultures were adjusted to OD = 0,6. For degradation experiments, a stock solution of AFB1 (1000 ppm in acetone) was used to supplement cultures to a final concentration of 2 ppm. Degradation experiments were carried out in 50 ml LB medium. A microbe free blank (50 ml LB with 2ppm AFB1) and a negative control – inoculated with a non mycotoxin degrader *E. coli* K12 strain – were set in the experiment as well. Samples were taken from degradation systems in every 24 hours. These samples were centrifuged at 4600 rpm, 4°C, 20 min. The supernatant and pellet were separated, and stored at -20°C until examination. The supernatant and pellet were analysed by High Performance Liquid Chromatography (Wessling Hungary Ltd., Hungary) and by an ELISA test (Soft Flow Ltd., Hungary). We also examined the supernatant of degradation systems by SOS-Chromotest.

2.3.1 Enzyme linked immunosorbent assay (ELISA-tests)

Toxin concentrations were determined by ELISA kits by the use of the TOXIWATCH system, SoftFow Biotechnology Ltd, Hungary. Measurements were carried out according to the manufacturer's instructions. In Toxiwatch ELISA Kits methanol (9,4%) is used for toxin extraction, thus standard lines contain methanol as well. As the matrix of the bacterial supernatant was a LB medium that does not contain methanol, in the case of AFB1 standard points without methanol had to be used. Measurements were carried out in triplicate.

2.3.2 High performance liquid chromatography (HPLC)

Chemical analytical tests were made by Wessling Hungary Ltd. Toxin concentrations were determined by the standard ethanol method. After cleaning up immunoaffinity columns and derivatization, liquid chromatography for separation and fluorescence detection was used according to AOAC Official Method 990.33. Measurements were carried out with Agilent 1100 HPLC-FLD. The column was Supelco and the temperature of the column thermostat was

25°C. Detection wavelengths regarding AFB1 were 365 nm (extinction) and 450 nm (emission). Injection happened with 100, 50, 10 μl and the eluents were water (68%), acetonitrile (50%), isocratic.

2.3.3 Application of SOS-Chromotest for analysing AFB1 biodegradation

We have carried out the SOS-Chromotest as described above, except we have tested the supernatant of negative control in the biodegradation test – bacteria (*E. coli*) without mycotoxin degrading ability – here as a positive genotoxic control. 10 μl of supernatant originated from degradation systems, a sample of controls and blank were added to corresponding wells and tested by SOS-Chromotest. We have compared the results of the SOS-Chromotest with parallel ELISA and chemical analytical tests.

2.3.4 Application of *A. fischeri* luminescence assay for analysing AFB1 biodegradation

We have classified microbes into three groups: microbes that have the highest, medium and insufficient AFB1 degradation potential. In that case we have set AFB1degradation systems with 4 ppm AFB1 in three parallels, where microbes were cultured in half strength (half amount of salts) Bactomarine broth. Moreover, we set negative controls without AFB1 contamination for clarifying whether the metabolic by-products of these microbes have a toxic affect on *A. fischeri*. In every 24 hours we took 8-8-8 ml samples from parallel systems and centrifuged them as described above. The supernatant was decanted and filtered through 0,45 μm membrane, then 10 ml of samples were distributed in 100 ml Erlenmayer flaks. We recovered salts in samples corresponding to the original Bactomarine broth, vortexed them, then filled them up with liquid culture of *A. fischeri*, in which the optical density was adjusted to 0,2 at 550 nm. Luminescence at t_0 was checked, then the samples were incubated at 25°C, 30 rpm in a shaking thermostat, and toxicity was determined as described above.

3 Results and discussion

3.1 Biomonitoring of mycotoxins with SOS-Chromotest

We have analysed AFB1 (10–0,078 μg/ml) in two fold serial dilution, and ZEA, OCHRA, T2 and DON in two concentration ranges (10 and 5 μg/ml). On the basis of our results, the SOS-Chromotest proved to be a less effective tool for screening the biological effects of ZEA, OCHRA, T2 and DON. In the case of these toxins the difference of genotoxic activity between native and activated mycotoxins were not significant; IF was 1,7±0,2 at 10 μg/ml and 1,5±0,2 at 5 μg/ml. In the case of AFB1 we found that as low as 0,078 μg/ml concentration of metabolically activated AFB1 induced the SOS repair system of the test bacterium and indicated a highly genotoxic 2,0 IF number (Fig 2.). These findings fit well with the published SOS-Chromotest results and underline the danger of AFB1 – the most potent carcinogenic natural substance. A detectable level of AFB1 approaches limits set in food and feed safety, thus the SOS-

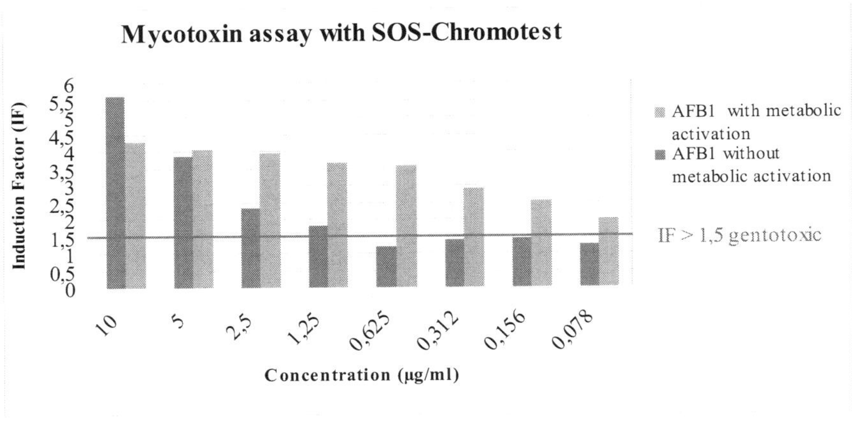

Figure 1: Genotoxicity of AFB1 measured by SOS-Chromotest.

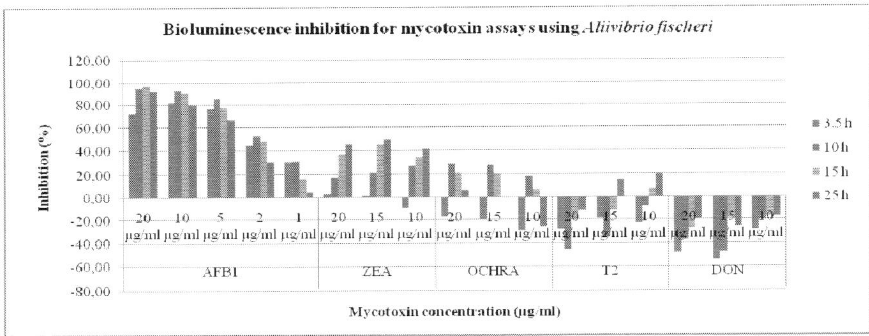

Figure 2: Bioluminescence inhibition for mycotoxins using *A. fischeri*.

Chromotest can be an appropriate tool for direct monitoring of AFB1 contaminated samples.

3.2 Biomonitoring of mycotoxins with *A. fischeri*

Our aim was to examine AFB1, ZEA, OCHRA, T2 and DON and to provide information about their toxicity with the help of an *A. fischeri* luminescence bioassay. Up to now only one publication dealt with this method oriented to mycotoxins [22], where AFB1 (10 µg/ml) and DON (20 µg/ml) were tested. We have tested three more toxins and all were used in a much wider concentration range. The *A. fischeri* luminescence assay showed great sensitivity in the case of AFB1, yet 1 µg/ml concentration of AFB1 was detectable. On the basis of our results, beside AFB1, ZEA has inhibited luminescence by 50% at 15 µg/ml range. Regarding DON, we measured similar luminescent gain in the manner of Sarter et al. [22]. However, adequate response in the case of T2 and OCHRA could not be measured. The results are shown in Fig. 2. According to our measurements this method is not suitable for direct monitoring, even of AFB1 contamination of food and feed stuffs, as it works at a much higher concentration than the current limits.

3.3 Application of SOS-Chromotest for analysing AFB1 biodegradation

We set AFB1 degradation tests to find microbes that have outstanding AFB1 degradation potential. We analysed the AFB1 degradation ability of 11 microbes on AFB1 (2 ppm) contaminated samples (Table 1). Our assumption was that the genotoxicity of samples ceased when microbes degraded AFB1 without forming a genotoxic by-product. The genotoxicity of the examined samples is expressed in Induction factors (IF). Out of 11 microbes we have selected six strains that successfully degraded AFB1 in 72 hours. Moreover, we have determined four strains that could eliminate genotoxic contamination in 48 hours. Our results and those made by ELISA-tests and by HPLC were highly correlated. In most cases genotoxicity ceased when ELISA-tests and HPLC indicated the decrease of AFB1 concentration. Interestingly, IFO 12538 showed genotoxicity, while great toxin elimination was detected (>90%) by ELISA and HPLC tests. In the case of this strain appearance, genotoxic by-products are assumable, which underlines the necessity of such biomonitoring measurements. Coherence between degradation ability and genotoxicity are in clear relief in the case of strains that have inefficient AFB1 degradation potential (N361, N58), which IF approximates the genotoxicity of negative control sample injected with *E. coli.* Furthermore, it indicates that different rhodococcus strains may use different routes for AFB1 degradation.

Table 1: Results of AFB1 (2ppm) biodegradation tests.

Microbe	Strain	SOS-Chromotest			ELISA-system	Chemical analysis
		Induction Factor			Degradation (%)	Degradation (%)
		24 h	48 h	72 h	72 h	72 h
E. coli	C	2.42	2.13	2.32	<20	<20
R. erythropolis	N11	1.98	1.36	1.3	97.64	81.71
	DSM4306	2.13	1.32	1.2	97.56	99.98
	NCIMB9784	2.09	1.82	1.2	97.11	98.55
	IFO12538	–	–	1.6	96.40	94.52
	OM72	3.09	2.1	1.68	79.20	84.30
	ZFM231	–	–	1.84	72.87	71.69
R. ruber	N361	–	–	3.0	<20	<20
R. globerulus	N58	–	–	2.6	20.79	<20
R. rhodochrous	CW25	1.83	1.26	1.2	97.97	99.98
	ATTC12674	2.3	1.4	1.2	97.90	96.05
	K402	–	1.92	1.38	98.11	99.92

3.4 Application of *A. fischeri* luminescence assay for analysing AFB1 biodegradation

Strains were classified in three groups on the basis of ELISA-tests and chemical analysis: outstanding (80–100%), good (70–80%) and insufficient AFB1 degrading strains. We analysed these strains by an *A. fischeri* luminescence inhibition test for the remaining toxicity of their AFB1 biodegradation. The results of the inhibition test are illustrated in Figure 3. High luminescence inhibition was found in the case of "insufficient" rhodococcus strains (N58, N361). "Good" strains (ZFM231, OM72) showed less inhibition than insufficient strains, but caused greater inhibition than excellent strains. Those strains that have the greatest, more than 99%, degradation potential did not cause inhibition for the third day of the biodegradation process, rather increased luminescence. Interestingly, in the *A. fischeri* luminescence assay, HPLC and ELISA-tests showed great degradation ability for *R. erythropolis* IFO11453, but at the same time the SOS-Chromotest showed genotoxicity. In the case of ATTC12674 slight cytotoxic effect was demonstrable after great toxin degradation. This underlines the necessity of using different bacterial biotests, since these tests measure different biological effects: genotoxicity and cytotoxicity.

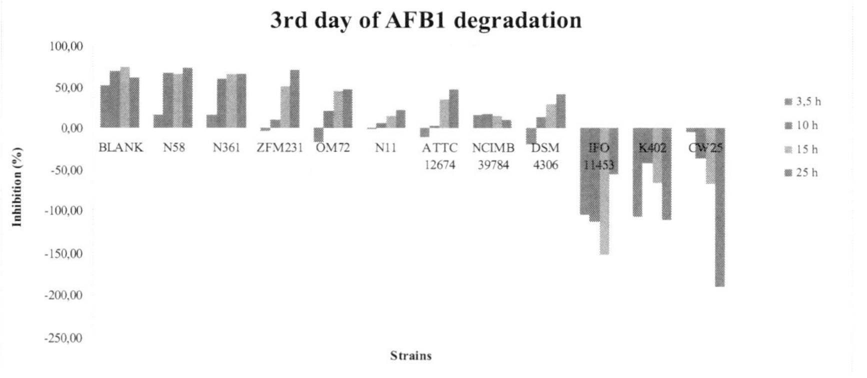

Figure 3: Luminescence inhibition by AFB1 biodegradation derivatives on *A. fischeri*.

4 Conclusion

We have adopted two bacterial biotests, *A. fischeri* luminescence assay and SOS-Chromotest, for monitoring AFB1, ZEA, OCHRA, T2 and DON contamination. The toxin levels of ZEA, DON, OCHRA and T2 needed for an SOS-system or an *A. fischeri* assay were two or three order of magnitude higher than current mycotoxin limits – so direct application of these tests is not possible. However, SOS-Chromotests would be a suitable tool for AFB1 monitoring, since the detectable limit of AFB1 by this test is around the regulated maximum levels of

AFB1 contamination in food and feedstuffs. Nevertheless, both methods seem to be suitable for screening mycotoxin degrading microbes. The SOS-Chromotest is suitable for monitoring biodegradation purposes of AFB1, as the test is appropriate for detecting the most dangerous biological effect of aflatoxin. *A. fischeri* luminescence assay is also suitable for screening AFB1 degrading microbes, since the *A. fischeri* test showed slighter toxicity in the case of strains with the best degradation ability in correllation with the paralell analitical and ELISA methods. In particular, the simultaneous application of both biotests was useful, as it cleared up any possible remaining mutagen effect even in the case of microbes with excellent degradation potential. Moreover, in the *A. fischeri* system a remaining cytotoxic effect was demonstrable after great toxin degradation and ceased genotoxicity. Thus, the combination of the above mentioned tests are crutial for the selection of safe and active microbial strains for further biodetoxification technologies. To sum up, the *A. fischeri* luminescence assay and SOS-Chromotest are definitely applicable for screening microbial strain collections, thousands of genetically improved microbe clones or transposon mutagenesis libraries for AFB1 degradation activity, as these are reliable, relatively simple, cheap and rapid methods.

Acknowledgement

This study was supported by the NKTH TECH_08-A3/2-2008-0385 (OM-00234/2008) MYCOSTOP grant.

References

[1] Enomoto, M., Saito, M., 1972. Carcinogens produced by fungi. Annual Review of Microbiology 26: 279–312.

[2] Sarter, S., Zakhia, N., 2004. Chemiluminescent and bioluminescent assays as innovative prospects for mycotoxin determination in food and feed. Luminescence 19: 345–351.

[3] Fernandez-Alba, A.R., Guil, M.D.H., Lopez, G.D., Chisti, Y., 2002. Comparative evaluation of the effects of pesticides in acute toxicity luminescence bioassays. Analytica Chimica Acta 451: 195–202.

[4] Peinado, M.T., Mariscal, A., Carnero-Varo, M. Fernandez-Crehuet, J., 2002. Correlation of two bioluminescence and one fluorogenic bioassay for the detection of toxic chemicals. Ecotoxicology and Environmental Safety 53: 170–177.

[5] Repetto, G., Jos, A., Hazen, M.J., Molero, M.L., Del Peso, A., Salguero, M., Del Castillo, P., Rodriguez-Vicente, M.C. and Repetto, M., 2001. A test battery for ecotoxicological evaluation of pentachlorophenol. Toxicology in Vitro 15: 503–509.

[6] Ribo, J.M., Kaiser, K.L.E., 1983. Effects of selected chemicals to photoluminescent bacteria and their correlation with acute and sublethal effects on other organisms. Chemosphere 12: 1421-1442.

[7] Kaiser, K. L. E., 1998. Correlation of *Vibrio fischeri* Bacteria Test with Bioassay Data for Other Organisms. Environmental Health Perspectives 106: 583–591

[8] Steinberg, S.M., Poziomek, E.J., Engelmann, W.H., Rogers, K.R., 1995. A review of environmental applications of bioluminescence measurements. Chemosphere 30(11): 2155–2197

[9] Cronin, M.T.D., 1999. Microtox as a Substitute for Acute Fish Toxicity Testing. TestSmart – A Humane and Efficient Approach to SIDS Data Workshop of the Johns Hopkins Center for Alternatives to Animal Testing, April 26-27 Fair Lakes, Fairfax

[10] Farre M, Barcelo D., 2003. Toxicity testing of wastewater and sewage sludge by biosensors, bioassays and chemical analysis. Trends Anal Chem; 22(5): 299–310.

[11] Quillardet, P., Hofnung, M., 1985. The SOS Chromotest, a colorimetric bacterial assay for genotoxins: procedures. Mutation Research 147(3):65-78

[12] Quillardet, P., de Bellecombe, C., Hofnung, M., 1985. The SOS Chromotest, a colorimetric bacterial assay for genotoxins: validation study with 83 compounds. Mutation Research 147(3): 79–95.

[13] Auffray, Y., Boutibonnes, P., 1984.Genotoxic activity of some mycotoxins using the sos chromotest. Mycopathologia 100: 49–53.

[14] Mamber, S. W., W. G. Okasinski, C. D. Pinter, and J. B. Tunac. 1986. The Escherichia coli K-12 SOS Chromotest agar spot test for simple, rapid detection of genotoxic agents. Mutat. Res. 171:83–90.

[15] Brams A, Buchet JP, Crutzen-Fayt MC, De Meester C, Lauwerys R, Léonard, A., 1987. A comparative study, with 40 chemicals, of the efficiency of the Salmonella assay and the SOS chromotest (kit procedure). Toxicol Lett. 38: 123–133

[16] von der Hude W, Behm C, Gürtler R, Basler A., 1988. Evaluation of the SOS chromotest. Mutat Res. 203(2):81–94

[17] Eder E; Deininger C; Kütt W, 1989. Genotoxicity of monofunctional methanesulphonates in the SOS chromotest as a function of alkylation mechanisms. A comparison with the mutagenicity in S. typhimurium TA100. Mutation research 211: 51–64

[18] McDaniels, A. E., Reyers, A. L., Wymer, L. J., Rankin, C. C., and Stelma, G. N., 1990. Comparison of the Salmonella (Ames) test, umu test, and the SOS chromotest for detecting genotoxins. Environ. Mol. Mutagen. 16: 204–215.

[19] Xu, H.H. and Schurr, K.M. (1990) Genotoxicity of 22 pesticides in microtitration SOS chromotest, Toxicity Assess., 5: 1–14

[20] Quillardet, P., Huisman, O., D'ari, R., Hofnung, M., 1982. SOS chromotest, a direct assay of induction of an SOS function in Escherichia coli K-12 to measure genotoxicity. Proc. Natl. Acad. Sci. USA Genetics 79: 5971-5975.

[21] Legault, R., Blaise, C., Rokosh, D., Chong-Kit, R., 1994. Comparative assessment of the SOS Chromotest kit and the Mutatox test with the Salmonella plate incorporation (Ames test) and fluctuation tests for

screening genotoxic agents. Environmental Toxicology and Water Quality 9: 45-57.

[22] Sarter, S., Metayer, I., Zakhia, N., 2008. Effects of mycotoxins, aflatoxin B1 and deoxynivalenol, on the bioluminescence of *Vibrio fischeri*. World Mycotoxin Journal 1(2): 189-193.

[23] Froehner, K., Meyer, W., Grimme, L.H., 2002. Time-dependent toxicity in the long-term inhibition assay with *Vibrio fischeri*. Chemosphere 46: 987-997.

[24] Teniola, O.D., Addo, P.A., Brost, I.M., Färber, P., Jany, K.-D., Alberts, J.F., Van Zyl, W.H., Steyn, P.S., Holzapfel, W.H., 2005. Degradation of aflatoxin B1 by cell-free extracts of *Rhodococcus erythropolis* and *Mycobacterium fluoranthenivorans* sp. nov. Int. J. Food Microbiol. 105: 111–117.

[25] Alberts, J.F., Engelbrecht, Y., Steyn, P. S., Holzapfel, W.H. and van Zyl W., H. (2006): Biological degradation of aflatoxin B1 by *Rhodococcus erythropolis* cultures. International Journal of Food Microbiology 109, 121–126

Role of fulvic acid on the reduction of cadmium toxicity on tilapia (*Oreochromis niloticus*)

A. E. Noor El Deen, M. S. Zaki & H. A. Osman
Hydrobiology Department, Veterinary Division,
National Research Centre, Egypt

Abstract

The effect of fulvic acid on cadmium (Cd) toxicity, the impact on fish immunological, and haematological changes in Nile tilapia (*Oreochromius niloticus*) were studied. The fish (100±10g) were exposed to 10 ppm Cd alone or with 0.1, 0.2 and 0.3 ppm for 15 and 45 days. Cd exposure reduced significantly ($P<0.04$), for example the erythrocyte count (RBCs), haemoglobin content (Hb), haematocrit value (Hct), mean cell haemoglobin (MCH) and mean cell haemoglobin concentration. These parameters were improved when fulvic acid was applied with Cd. The values of RBCs, Hb, Hct, MCH and MCHC were increased significantly in the control fish group. The addition of fulvic acid to Cd contaminated medium considerably reduced metal absorption and accumulation in fish tissues, while metals in water and faeces increased. Fish exposed to Cd alone accumulated 2.15 and 5.970 mg Cd/g dry weight in the liver tissue over 15 and 45 days, respectively. Cd reduced significantly to 1.292 and 4.16.; 0.92 and 3.791; and 0.41 and 2.43 mg Cd/g dry weight tissue in fish exposed to 0.1, 0.2 and 0.3g fulvic acid/L over 15 and 45 days, respectively. Similar trends were observed in gills and musculature.
Keywords: Nile tilapia, cadmium, immunological, fulvic acid, haematology, liver, gills, musculature.

1 Introduction

Nile tilapia are considered the most popular widely distributed, cheapest and intensively cultured fish in Egypt. The clinical picture in naturally infested and polluted Tilapia sp was revealed, some were aggregated on the water surface, and accumulated at the water inlet of the pond and air pump of aquaria. Almost all appeared dull with a loss of escape reflex (Eissa et al. [1] and Eaton and

WIT Transactions on Ecology and the Environment, Vol 132, © 2010 WIT Press
www.witpress.com, ISSN 1743-3541 (on-line)
doi:10.2495/ETOX100151

Stinson [2]). The present study reveals that the fish exposed to Cd alone showed significant reduction in RBCs, Hb and Hct, compared with those exposed to Cd with different levels of fulvic acid. The reduction of these parameters in Nile tilapia, O niloticus, at sublethal levels of cadmium might be due to the destruction of mature RBCs and the inhibition of erythrocyte production due to reduction of haemsynthesis, which is affected by pollutants (James and Sampath [3]). Also, the decrease in the RBC count may be attributed to haematopathology or acute haemolytic crisis that results in severe anaemia in most vertebrates, including fish species exposed to different environmental pollutants (Yamawaki et al. [4]) or it may be that the decrease in RBCs can attributed to the reduction of growth and other food utilization parameters, which results in sever anaemia (Wintrobe [5]). Moussa [6] found a significant decreased in total erythrocyte count, haemoglobin content, haematocrit value and mean corpuscular haemoglobin concentration in air breathing fish, Channa punctatus, after exposure to a sublethal dose of Cd (29 mg Cd/L). The addition of fulvic acid improves the haematological parameters (RBCs, Hb and Hct), which indicates the capability of fulvic acid to chelate Cd from the media. Subsequently, the Cd toxicity was reduced. These results are in agreement with those of Snedecor and Cochran [7], who observed that *Oreochromis mossambicus* exposed to copper along with fulvic acid showed a significant improvement in blood parameters over those of copper alone. The calculated blood indices MCV, MCH and MCHC have a particular importance in anaemia diagnosis in most animals (Coles, 1986). The perturbations in these blood indices (increased MCV, decreased MCH and MCHC) may be attributed to a defence against Cd toxicity through the stimulation of erythropiosis or may be related to the decrease in RBCs, Hb and Hct due to the exaggerated disturbances that occurred in both metabolic and hemopoietic activities of fish exposed to sublethal concentration of pollutants (Huang et al. [8]). The present results indicate that fulvic acid is effective in removing Cd from water, and reducing Cd bioaccumulation in fish. Particulate organic matter can scavenge metal from water and help to reduce metal from fish. These results are in agreement with Shalaby [9], whose study shows that any agent that can remove Cd from water helps to reduce the bioaccumulation of this metal in fish. The present study showed that the addition of fulvic acid to the Cd media reduced significantly ($P<0.05$) the Cd level in water and metal uptake as compared to fish exposed to Cd alone. The Cd concentration in water was 9.31 mg/L and it decreased significantly ($P< 0.05$). The Cd accumulation in the liver, gills and muscle of fish exposed to Cd alone was higher than that of fulvic acid. These results suggest that fulvic acid could chelate Cd ions producing a stable complex, thus reducing the chance for metal uptake by tissues. Besides, the fulvic acid eliminated a greater amount of Cd from the body through faeces. The formation of a Cd-fulvic acid complex in water and the elimination of a greater amount of Cd in faeces evidently reduced the metal burden in tissues and thereby improved the haematological parameters of fish exposed to Cd. Planas-Bohne and Lehman [10] found a low level of cadmium in tissues due to increased excretion of metals through faeces and urine when rats were administered Cd intravenously along with fulvic acid (Table 1).

www.witpress.com, ISSN 1743-3541 (on-line)

Table 1: Changes in cadmium residue in water (mg Cd/L), liver, gills, musculature and faeces (mg Cd/g dry weigh) of Nile tilapia (O. niloticus) exposed to Cd with or without fulvic acid (FA).

Items	Water	Liver		Gills		Musculureate		Faeces	
Period/days		15	45	15	45	15	45	15	45
Control	0.041	0.048^{a}	0.055^{a}	0.038^{a}	0.039^{a}	0.023^{a}	0.076^{a}	0.003^{a}b	0.005^{a}b
Cd	± 0.02 9.31	± 0.02 2.15^{b}	± 0.004 5.971^{b}	± 0.02 1.36^{b}	± 0.04 2.56^{b}	± 0.002 0.476^{b}	± 0.005 1.077^{b}	±0.018 0.153^{b}	± 0.02 0.189^{b}
Cd+0.1g	± 0832 7.15	± 0.253 1.292^{c}	± 0.85 4.16^{b}	± 0.085 0.65^{c}	± 0.276 1.07^{c}	± 0.06 0.343^{cb}	± 0.15 0.665^{c}	± 0.018 0.940^{c}	± 006 2.067^{c}
FA /1 Cd+0.2g	± 0.34 3.78	± 0.056 0.95^{d}	± 0.45 3.791^{b}	± 0.06 0.394^{d}	± 0.11 0.85^{c}	± 0.04 0.33^{cb}	± 0.021 0.383^{d}	± 0.03 2.34^{d}	± 0.143 5.443^{d}
FA/1 Cd+0.3g	± 0.01 1.73	± 0.054 0.42^{e}	± 0.29 2.45^{c}	± 0.052 0.266^{d}	± 0.06 0.71^{c}	± 0.08 0.216^{c}	± 0.034 0.217^{d}	± 0.069 5.282^{e}	± 0.345 7.456^{e}
FA/1	± 0.02	± 0.034	± 0.23	± 0.073	± 0.42	± 0.03	± 0.025	± 0.32	± 0.528

2 Polluted tilapia....why?

Cadmium is one of the most toxic heavy metals that enters the environment from natural sources and as a result of man's activity, such as recycling of scrap metal, electroplating, industry manufacturing vinyl plastics, electrical contacts, metallic and plastic pipes. Tilapia have the capability of concentrating metals by feeding and metabolic processes, which can lead to the accumulation of high concentrations of metals in their tissues. The reduction of toxic elements, such as cadmium, in aquatic environments is needed by any acceptable method.

2.1 Are there solutions?

The most widely used technique for the removal of toxic elements involves the process of neutralization and metal hydroxide precipitation (Hiemesh and Mahadevaswamy [11]). Chemicals can effectively remove certain toxic elements from industrial wastes or polluted media, but it is usually costly. However, there are some cheap natural products that are also free from undesirable side effects. In recent years, the remobilization of metals by synthetic anthropogenic chelating agents has received much attention. The literature reported a number of chelators that have been used for chelate-induced hyper accumulation (Khangarot and Tripathi [12]). Natural compounds, such as fulvic acid, are known to be effective chelating agents of heavy metals (Karuppasamy [13]).

2.2 Why fulvic acid?

Fulvic acid is the most commonly used chelator because of its small molecular weight and strong chelating ability for different heavy metals (Litchfield and

Wileoxon [14] and Donor and William [15]). Metal bioaccumulation can occur via complication, coordination, chelation, ion exchange and other processes of greater or lesser specificity. Bioaccumulation processes are sometimes due to active (metabolism dependent) metal accumulation by living cells. In spite of the amount of data published on the effect of water borne exposure of cadmium and fulvic acid singly, information on the effects of a Cd/fulvic acid mixture on aquatic organisms are limited and not uniform. Therefore, fulvic acid appears to be a promising tool to control cadmium pollution in aquaculture. The present study, short- and long-term bioassays were designed to evaluate the influence of fulvic acid on the retention of cadmium in water. It was carried out to investigate the effect of fulvic acid on the reduction of the toxicity of cadmium, to enhance the change of blood parameters and enzymes and to assess its impact on some physiological parameters of Nile tilapia *(Oreochromis niloticus)*.

3 Practical procedures

The present study showed that the addition of fulvic acid to Cd contaminated media reduced significantly the Cd level in water and helped to eliminate Cd from the fish body, which in turn improved the clinical signs and the haematological parameters as compared to fish exposed to cadmium alone.

3.1 Collecting sample tilapia

A healthy 75 fish of Nile tilapia *Oreochromis niloticus* weighing (100±10g)/fish were collected from the ponds of Kafr Eel Sheikh governorate fish farms, Egypt. Fish were acclimated in cement fish ponds for two weeks. Acclimated fish were exposed to different concentrations of cadmium and mortality was observed for 96-h. A static renewable bioassay method (Duncan [16]) was adopted for the determination of 96-h median lethal probity analysis; Santschi [17] was followed for the calculation of 96 hr LC_{50}. A control group was maintained in metal-free tap water. The 96 hr LC_{50} of cadmium for *Oreochromis niloticus* was 40 ppm. A stock solution of cadmium was prepared by dissolving 10.686 g of annular grad cadmium sulphate ($CdSO_4$– $8/3H_2O$) in 1/L of distilled water and diluted with water to obtain the desired concentration (10 ppm) for this experiment. The fish were distributed randomly in five cement ponds at a rate of 15 fish/aquarium, which containing aerated tap water. These aquaria were divided into five groups with three replicates each per group. Fish were fed frequently on a diet containing 25% crude protein (CP) at a rate of 3% of live body weight twice daily for 15 and 45 days. Siphoning of three quarters of the aquariums was done every day for waste removal and it was replaced by an equal volume of water containing the same concentration of Cd and fulvic acid. Dead fish were removed and recorded daily.

3.2 Sample classification

The first group was free of Cd and fulvic acid and maintained as a control. The second group was exposed to 10 ppm of Cd SO_4 only. The third, fourth and fifth groups were exposed to 10 mg Cd/L and 0.1, 0.2 and 0.3 g fulvic acid/L, respectively. Each aquarium was supplied with compressed air via air-stones from air pumps. A well-aerated water supply was provided from a storage fibreglass tank. The temperature was adjusted at 27±1°C by means of thermostats (Table 2).

Table 2: Field experimental groups and their notation.

S. No.	Groups in field ponds	Nation
1	Control (metal free water)	C
2	Cadmium (10 ppm) alone	Cd
3	Cadmium (10 ppm) +0.1g fulvic acid /l	Cd fulvic acid 1
4	Cadmium (10 ppm) +0.2 fulvic acid /l	Cd fulvic acid 2
5	Cadmium (10 ppm) +0.3g fulvic acid /l	Cd fulvic acid 3

3.3 Cd residue

Cadmium sulphate and fulvic acid was obtained from the El-Nasr chemical and Grotech companies (Egypt), respectively, and prepared in aquatic solution to provide the required concentrations of cadmium and fulvic acid. Cadmium was measured in the water, liver, gills, musculature and faeces according to method of Norvell [18].

3.4 Statistical analysis

The obtained data were subjected to analysis of variance between means and were done at the 5% probability level, using Duncan's new multiple range test by Spraggue [19].

4 Results

The present study showed that addition of fulvic acid to Cd contaminated media reduced significantly the Cd level in the water and helped to eliminate metal from the fish body (liver, gills, musculature and faeces) and in turn improved the biochemical parameters as compared to fish exposed to Cd alone (see Table 2).

4.1 Clinical examination

The clinical examination of most examined fish showed asphyxia, some aggregated on the surface, accumulated at the water inlet of the pond and the air pump of aquaria. Others appeared dull with loss of escape reflex.

4.1.1 Haematological parameters

The results of erythrocyte count (RBCs), haemoglobin content (Hb) and haematocrit value (Hct) obtained from the fish exposed to a sublethal dose of Cd

Table 3: Changes in mean cell volume (MCV), mean cell haemoglobin (MCH) and mean cell haemoglobin concentration (MCHC) in the blood of Nile tilapia (O. niloticus) exposed to Cd with or without fulvic acid (FA).

Items	MCV		MCH		MCHC	
Period	15 days	45 days	15 days	45 days	15 days	45 days
Control	95.32 ^{ad}	100.02 ^{a}	34.35 ^{a}	43.21 ^{a}	34.77 ^{a}	43.32 ^{a}
	± 1.86	± 2.243	± 0.342	± 1.432	± 1.121	± 0.928
Cd	106.93 ^{b}	106.75 ^{b}	33.02 ^{b}	36.56 ^{b}	31.15 ^{b}	37.27 ^{b}
	± 2.23	± 0.874	± 0.177	± 0.846	± 0.909	± 1.85
Cd+0.1 g	93.45 ^{a}	95.71 ^{a}	34.52 ^{a}	32.45 ^{b}	37.21 ^{a}	33..77 ^{b}
FA/l	± 2.05	± 4.26	± 1.23	± 1.17	± 1.26	± 1.49
Cd+0.2 g	96.24 ^{a}	98.77 ^{a}	33.23 ^{a}	36.76 ^{bc}	34.06 ^{ac}	37.40 ^{b}
FA/l	± 2.64	± 0.909	± 1.72	± 1.49	± 1.76	± 1.20
Cd+0.3 g	101.1 ^{db}	107.95 ^{b}	32.85 ^{a}	37.02 ^{ac}	33.34 ^{cb}	34.87 ^{b}
FA	± 2.512	± 2.241	± 1.702	± 1.576	± 0.941	± 1.68

Table 4: Changes in erythrocyte (count x 106/mm3), haemoglobin content (g/100ml) and haematocrit value (%) in the blood of Nile tilapia (O. niloticus) exposed to Cd with and without fulvic acid (FA).

Items	Erythrocyte count (RBCs)		Haemoglobin (HB)		Haematocrit value (Hct)	
Period	15 days	45 days	15 days	45 days	15 days	45 days
Control	1.58 ^{a}	1.714 ^{a}	5.48 ^{a}	7.315 ^{a}	15..31 ^{a}	17..32 ^{a}
	± 0.073	± 0.051	± 0.353	± 0.133	± 0.308	± 1.665
Cd	1.268 ^{b}	1.06 ^{b}	4.21 ^{b}	4.12 ^{c}	13..5 ^{b}	12.01 ^{b}
	± 0.064	± 0.073	± 0.235	± 0.354	± 0.47	± 0.576
Cd+0.1g	1.572 ^{a}	1.57 ^{d}	4.54 ^{ab}	5.12 ^{b}	14.66 ^{a}	15.05 ^{a}
FA/l	± 0.064	± 0.023	± 0.395	± 0.136	± 1.454	± 0.76
Cd+0.2g	1.56 ^{a}	1.786 ^{ac}	5.17 ^{ab}	6.605 ^{b}	15.02 ^{a}	17.65 ^{a}
EDTA/l	± 0.086	± 0.032	± 0.458	± 0.305	± 1.72	± 0.916
Cd+0.3g	1.956 ^{c}	2.01 ^{c}	6.464 ^{c}	7.68 ^{a}	20.0 ^{c}	22.02 ^{c}
FA/l	± 0.086	± 0.063	± 0.277	± 0.133	± 0.365	± 1.471

(10 mg/l) alone or with different doses of fulvic acid are given in Table 4. Table 3 shows that the RBCs, HB and HCt were reduced in fish exposed to Cd over both periods and they were less than that of the control ($P<0.05$) The RBCs count decreased significantly in fish exposed to Cd at 15 and 45 days. On the other hand, these parameters returned to the normal values and increased significantly in fish exposed to Cd with 0.2 and 0.3 g of fulvic acid/L for 15 and 45 days. These values increased significantly in fish exposed to Cd with 0.3 g fulvic acid/L. Blood parameter were improved in fish exposed to Cd with different levels of fulvic acid. The blood indices calculated from the mean values of blood parameters for the aforementioned treatments are given in Table 3.

Data shows that the MCV increased significantly in fish exposed to Cd alone, while the MCH and MCHC decreased significantly in fish exposed to Cd only when compared with the control. These parameters increased with increasing of exposure time of fish to Cd. The addition of fulvic acid to Cd-polluted media maintained the MCV, MCH and MCHC at levels close to those of the control.

5 Recommendation

From the present study, it is recommended that an optimum dosage of 0.3 g fulvic acid/L can effectively chelate Cd from contaminated water. Hence, a scientific method of detoxification is essential to improve the health of fish in any stressed environmental conditions.

References

[1] Eissa, I.A.M and Fatma. M. Fakhry (1994): Clinical picture and residue levels of locally used pesticides (field dose) in some freshwater and marine fishes. J. Egypt Vet. Med. Res., 3, 1, 96-107.

[2] Eaton, D. L and Stinson, M. D. (1983): Concentration of lead, cadmium, mercury and copper in the cray fish (*Pacifasticus leniusculus*) obtained from a lake receiving urban runoff. Arch. Environ. Contam. Toxicol., 12: 693- 700.

[3] James, R. and Sampath, K (1999): Effect of the ion- exchanging agent, Zeolite, on reduction of cadmium toxicity: an experimental study on growth and elemental uptake in *Heteropneustes fossilis* (Bloch). J. Aqua. Trop., 14 (1) 65- 74.

[4] Yamawaki, K.; Hashimoto, W.; Fujii, K.; Koyama, J.; Ikeda, Y. and Ozaki, H. (1986): Hematological changes in carp exposed to low cadmium concentration. Bull of the Japanese. Soc. Sci. Fish., 59 (3):459- 466.

[5] Wintrobe, M. M (1978): Clinical hematology. Henry Kimpton, London, pp: 448.

[6] Moussa, M. A. (1999): Biological and physiological studies on the effect of the gramoxon and stomp herbicides on Nile tilapia *(Oreochromis niloticus*). Fact. Sci. Zool. Dep. Cairo. Univ.,200p (Ph.D. Thesis)

[7] Snedecor, G. W. and Cochran, W. G. (1982): Statistical Methods. 6*th* edition. Iowa State Univ. Press., Amer., IA, USA, pp 593.

[8] Huang, J. W.; Chen, J.; Berti, W. R. and Cunningham S. D (1997): Phytoremediation of Pb-contaminated soils: Role of synthetic chelates in lead phytoextraction. Environ. Sci. Technol., 31: 800- 805.

[9] Shalaby, A. M. (2001): Protective effect of ascorbic acid against mercury intoxication in Nile tilapia (*Oreochromius niloticus*). J. Egypt. Acad. Soc. Environ. Develop., (D- Environmental studies), 2 (3):79-97.

[10] Planas-Bohne, F and Lehman, M (1983): Influence of chelating agent on the distribution and excretion of cadmium in rats. Toxicol. Appl. Pharmacol., 67: 408- 416.

[11] Hiemesh, S. and Mahadevaswamy, M. (1994): Sorption potential of biosorban: for the removal of copper. Indian. J. Environ. Hlth., 36: 165-169.

[12] Khangarot, B. S and Tripathi, D. M (1991): Changes in humoral and cell-mediated immune responses and in skin and respiratory surfaces of cat fish *Saccobranchus fossillis*, following copper exposure. Ecot. Envir. Safety, 22 (3): 291- 308.

[13] Karuppasamy, R.; Subathra, S and Puvaneswari, S. (2005): Haematological responses to exposure to sublethal concentration of cadmium in air breathing fish, *Channa punctatus* (Bloch). J. Environ Biol., 26(1):123-8.

[14] Litchfield, J. T and Wileoxon, F. (1949): A simplified method for evaluating dose- effect experiments. J. Pharmacol. . Exp. Ther., 96:59-113.

[15] Donor and William R. (1993). Humic, Fulvic and Microbial Balance: Organic Soil Conditioning. Evergreen Colorado: Jackson Research Center.

[16] Duncan, D.B. (1955): Multiple range and multiple (F) test. Biometrics, 11: 1- 42.

[17] Santschi, P. H (1988): Factors controlling the biogeochemical cycle of trace elements in fresh and coastal waters as revealed by artificial radioisotopes. Limnology and oceanography, 33:848- 886.

[18] Norvell, W.A (1991): Reactions of metal chelates in soil and nutrient solutions In: Mortvedt, J.J., Cox, F.R., Shuman, L. M and Welch, R. M., Editors, 1991. Micronutrients in Agriculture, 2nd Edition, Soil. Science of America, Madison, Wisconnsin, pp. 187- 227.

[19] Spraggue, J.B (1973): The ABCs of pollutant bioassay using fish In Biological methods for Assessment of water quality. ASTM 528. Amer.testing. Material, pp.6-30.

Section 4
New trends in environmental toxicology

Technical issues surrounding the preparation, characterisation and testing of nanoparticles for ecotoxicological studies

R. Tantra, S. Jing & D. Gohil
National Physical Laboratory, UK

Abstract

There is a genuine concern about how engineered nanoparticles affect the environment and this has resulted in a detailed two-part study that is presented in this paper. The first part of the study investigates some of the issues surrounding the dispersion and characterisation of nanoparticle suspensions, which are critical in order to carry out testing for understanding the ecotoxicological properties of nanoparticles in the environment. Cerium oxide (CeO_2) nanoparticles were dispersed in de-ionised (DI) water and subsequently characterised using Dynamic Light Scattering and Scanning Electron Microscopy. Results showed that the reliability of data obtained depended heavily on the need to control the dispersion step and to understand limitations associated with current measurement techniques. The second part of the study investigated the fate of nanoparticles when dispersed in three different ecotox media (seawater compared with media of fish and Daphnia), in an attempt to identify initial measurement concerns. Visual sedimentation experiments showed that nanoparticles (within a two day period) were relatively unstable in these ecotox media (relative if dispersion was carried out in DI water). Although most particles aggregated into larger clusters, SEM images showed the presence of nanosize clusters (<800 nm), which were still present in these media. It is the presence of these nanosize particles that will be of utmost concern, if the hypothesis that relates particle size and toxicological activity holds true.

Keywords: nanoparticles, aquatic ecotoxicity, characterisation, dispersion, aggregation, agglomerates, DLS, SEM.

WIT Transactions on Ecology and the Environment, Vol 132, © 2010 WIT Press
www.witpress.com, ISSN 1743-3541 (on-line)
doi:10.2495/ETOX100161

1 Introduction

Over the past few years, research concerning nanoparticle toxicity has attracted public concern [1]. Particularly, in assessing their toxicological significance, several studies [2, 3] indicated that nanoparticle toxicity was governed by their small size and high surface area, which subsequently would lead to greater chemical reactivity. In an attempt to address this public concern, the OECD (Organization for Economic Co-operation and Development) has recently launched a sponsorship program on nanoparticle safety assessment that requires global co-operation. This has resulted in the United Kingdom to launch the PROPEcT LINK project, which aims to fulfil the UK's contribution for the testing of zinc oxide (ZnO) and cerium oxide (CeO_2) nanoparticles.

Central to toxicological investigation of nanoparticle is the need to link toxicological activity with physicochemical properties [4]. In other words, what are the physical/chemical parameters of the nanoparticles that are most responsible for toxicological activity? This type of research has been conducted in the past but the conclusions drawn from such studies are often contradictory in nature, suggesting the need to successfully develop and establish internationally agreed standardised protocols. For example, the toxicity of carbon nanotubes has been one of the most pressing questions in nanotechnology [5]. Recently, Poland and co-workers [6] have shown experimentally that nanotubes display similar toxic responses to asbestos fibres, whereas Koyama et al. [7] have reported that the extent of toxicity of carbon nanotubes was low if compared to asbestos.

The purpose of the present study was to fulfil two objectives. The first objective concerns issues surrounding development of protocols. This research explores some of the issues associated with dispersing and subsequently characterising CeO_2 in DI water; DI water was used as past results showed good stability when nanoparticles were dispersed in such media as indicated by their high zeta-potential values [8, 9]. Results showing the effect of using different de-agglomeration tools to disperse the nanoparticles are presented in this paper. In the development of characterisation protocols, the importance of understanding the limitations of the technique (particularly "limit of quantification") for the intended use will be evaluated. For example, as nanotoxicological investigations often require the need to conduct analysis at extremely low particle concentrations, ~ nanogram per litre (ng/L) or less [10], the effects on data acquired (from various techniques: Dynamic Light Scattering (DLS) for particle size and zeta-potential measurement, and scanning electron microscopy (SEM)) upon dilution of nanoparticle concentration has been investigated. A technology "space map" is presented to show the limitations associated with using these tools.

The second objective of this work was to determine the fate of the nanoparticles in ecotox media and to identify any initial concerns. According to hypothesis, it is aspects of particle size characteristics that dominate the toxic profile of nanoparticles [11]. It would be expected that at high salt concentrations, the ecotox media would result in particle instability upon dispersion resulting in the formation of large aggregates/agglomerates that would eventually sediment out [12] and reduce toxicity. The central question here is: Will all of the nanoparticles sediment out? To assess this, dispersions of CeO_2 and ZnO were prepared in three different ecotox media (seawater, media of fish and Daphnia) and results from a visual sedimentation type experiment will be presented here. Also, results will be compared relative to dispersions in DI water. SEM analysis was used to characterise nanoparticles and agglomerates in dispersions after a two-day hold period.

2 Experimental

2.1 Materials and sample preparation

Z-Cote Zinc Oxide (ZnO with a nominal particle diameter size of 100 nm) and Nanograin (CeO_2, with an average particle size of ~50–70 nm) were supplied by BASF SE and Umicore Belgium, respectively. Nanoparticles were dispersed using the protocol below, in one of four possible aqueous liquid media: deionised (DI) water and three ecotox media (seawater and media of fish and Daphnia). DI water from Millipore, MilliQ system was used to prepare all aqueous solutions and suspensions. Ecotox media was prepared as follows: a) Seawater - 25 g per L of Tropic Marine Sea Salt (Tropical and Marine Limited), was prepared, pH ~7.5. b) Daphnia freshwater media. Salts (196 mg $CaCl_2{\cdot}2H_2O$, 82 mg $MgSO_4{\cdot}7H_2O$, 65 mg $NaHCO_3$, 0.002 mg Na_2SeO_3 (as obtained by appropriate dilutions of a 2 mg/ml stock solution) were dissolved in 1 L of DIwater. Upon continued stirring, DI water was further added so that the final pH ~7.5 and conductivity was between ~360–480 μS/cm. End volume ~1–1.5 L. c) Fish freshwater media. This was prepared in three separate steps. First, salts (11.76 g $CaCl_2{\cdot}2H2O$, 4.93 g $MgSO_4{\cdot}7H_2O$, 2.59 g $NaHCO_3$, 0.23 g KCl) were dissolved separately in 1L of DI water to make four separate stock solutions. Second, 25 mL of each salt stock solution was aliquoted into a clean bottle and diluted in DI water (made up to 1 L volume). Third, 200 ml of the stock solution from Step 2 was aliquoted and further diluted with DI water (made up to 1L volume). For long-term storage, these ecotox solutions were autoclaved and kept refrigerated until needed.

2.2 Nanoparticle dispersion in aqueous liquid media

Nanoparticle powders were weighed into small clean vials using an analytical mass balance. To disperse, a few drops of the appropriate liquid media were added to the vial and mixed into a thick paste using a spatula. ~15 mL of liquid media was then added to the paste and the whole mixture gently stirred using a spatula. De-agglomeration step was then carried out and this is very much dependent on which de-agglomeration tools were employed. If an ultrasonic probe (130 Watt Ultrasonic Processors) was used, the ultrasonic probe tip (6 mm Ti) was inserted half way down the 15 ml volume of dispersed nanoparticles and sonicated with 90% amplitude for 20 s (unless otherwise stated); temperature measurements were made using a digital thermometer (Fisher Scientific) before and after the sonication step. After sonication, the nanoparticle suspension was diluted using the appropriate liquid media, in order to make up to 1 L total volume (unless otherwise stated), a glass rod was used to gently mix the final dispersion, to ensure homogeneity. If other de-agglomeration tools were used instead (PowerGen Fisherbrand 500 homogeniser or Kinematica Overhead stirrer PX-SR 90 D), then the initial 15 ml dispersion mixture was diluted straightaway into the appropriate liquid media, to make up a total 1 L in volume. The homogeniser or overhead stirrer was lowered into the dispersion and operated at a constant speed for 1 minute to create maximum vortex action without spillage in a 1 L beaker.

For the purpose of investigating "limit of quantification", CeO_2 was dispersed in DI water using an ultrasonic probe. A stock solution of 500 mg/L was prepared and appropriate dilutions with DI water were made from this stock. For the purpose of "visual sedimentation" tests, eight separate nanoparticle suspensions (500 mg/L) were prepared (ZnO and CeO_2 dispersed separately in DI water and three ecotox media) in media bottles. Optical images showing the state of the dispersion in the bottles were recorded using a digital camera at set intervals over a period of two days. In between recording the images, the bottles were stored in the dark.

2.3 DLS (particle size and zeta-potential) analysis

The instrument employed for particle size analysis and zeta-potential measurements was a Zetasizer Nano ZS (Malvern Instruments, UK) with 633 nm red laser. The same instrument is also able to make zeta-potential measurements, by using a laser Doppler electrophoresis configuration. Detailed protocols for DLS particle size and zeta-potential measurements have been reported elsewhere [8].

2.4 Scanning electron microscopy (SEM) analysis

Scanning electron microscope images were recorded using a Carl Zeiss Supra 40 field emission scanning electron microscope, in which the optimal spatial resolution of the microscope was a few nanometres. In-Lens detector images were acquired at an accelerating voltage of 15 kV, working distance of ~3 mm, tilt angle 0°. For analysis of the "as received" nanoparticle powder, a small scoop of the nanoparticle powder was sprinkled over an SEM carbon adhesive discs; one side of the carbon disc was placed securely on a metal stub, whilst the other side was exposed to the nanoparticle powder. Excess powder on top of the disc was removed by gently tapping the stub on its side until an even (light) coating of powder on the surface was apparent. Detailed protocols associated with sample preparation (of depositing nanoparticles dispersed in liquid media on to poly-L-lysine slides) suitable for SEM analysis have been reported elsewhere [8].

3 Results and discussion

3.1 Nanoparticle dispersion

SEM images of the "as received" for ZnO and CeO_2 nanoparticles are given in Figure 1a) and show that polydispersity for both particle size and shape is high. These SEM images also show evidence of extensive aggregation and agglomeration (fusion of particles) that exists in both nanoparticles; this is particularly evident in CeO_2.

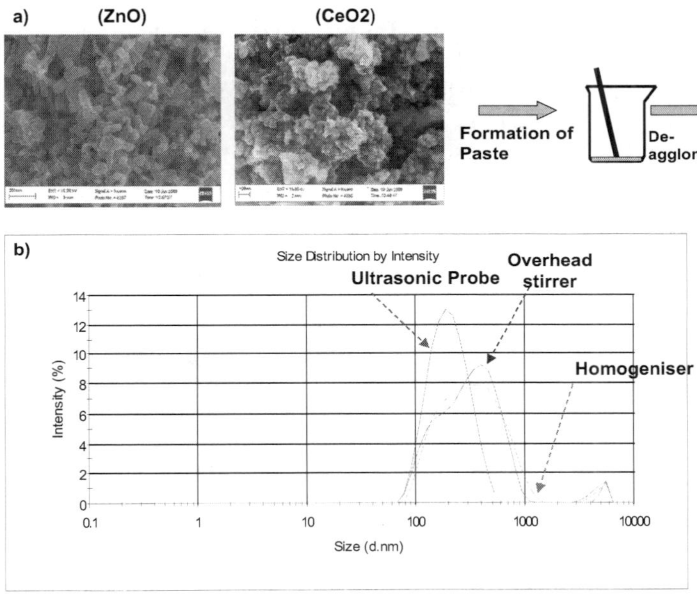

Figure 1: Nanoparticle dispersion in aqueous media: a) a schematic of the dispersion step from the "as received" powders (SEM images shown; scale bars 200nm for ZnO and 100 nm for CeO_2) b) Particle size distribution of CeO_2 in DI water (50 mg/L) and the effects of using different de-agglomeration tools at an exposure time of 1 minute).

Figure 1a) shows a schematic illustrating the steps of the dispersion protocol, as detailed in the Experimental section. Overall, this involved two essentials steps: a) the wetting of the nanoparticle powder into a paste, so as to substitute solid air-interfaces with solid-liquid interfaces, as recommended by ISO 14887: 2000 [13] b) de-agglomeration of nanoparticles using an appropriate tool, so as to introduce sufficient shear energy such that aggregates/agglomerates are broken down using an appropriate de-agglomeration tool, ideally to individual primary particles [14]. Figure 1b) shows the particle size distribution (by intensity as reported by DLS) of CeO_2 (50 mg/L) in DI water, when dispersed using an ultrasonic probe, with an exposure time of 1 minute. Results show a particle size distribution between 68–615 nm in size. The plot also shows the effect of altering the dispersion protocol step, when either an overhead stirrer or homogeniser was employed instead of ultrasonic probe. Results indicate a much broader particle size distribution, with particle sizes as big as 1 μm; the much bigger size particles found in the dispersions using these tools can only be explained by the much lower shear energy provided (to result in insufficient de-agglomeration/de-aggregation) if compared to the ultrasonic probe. Undoubtedly the final stability after the dispersion will be governed by the inherent properties of the liquid media and their interactions with the nanoparticles [15]. Overall, due to its effectiveness in de-agglomerating, the ultrasonic probe proved to be the tool of choice for the dispersion protocol.

Another variable that can potentially affect the particle size distribution is the length of time that the dispersion is exposed to i.e. the "exposure time". Figure 2a). shows the effect of changing the "exposure time" on the mean particle size. As expected, increasing the de-agglomeration time from 5 s to 50 s resulted in a reduction of particle size; the longer the exposure time the more de-agglomeration occurred resulting in smaller particle sizes. However, increasing

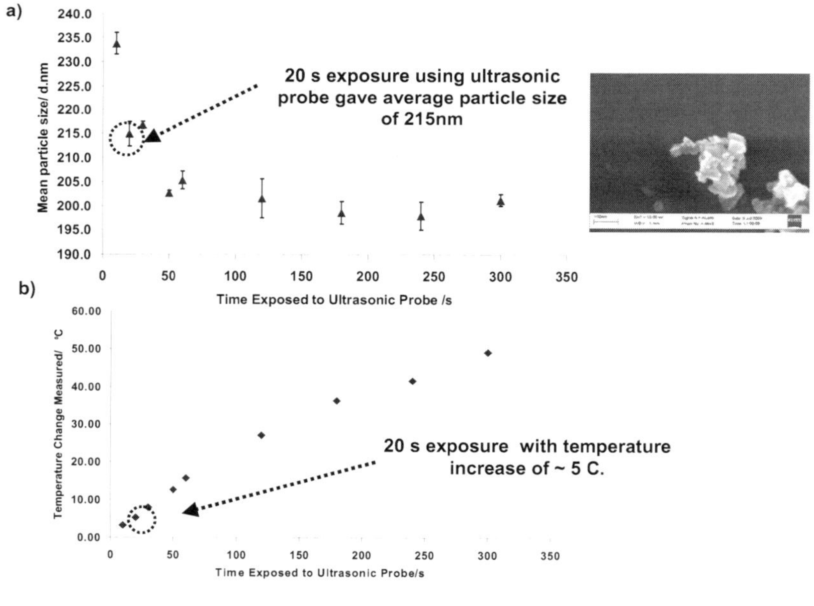

Figure 2: Effect of varying "exposure time" (using an ultrasonic probe) on the DLS mean particle size of CeO_2 in DI water (50 mg/L) on: a) mean particle size (inset: corresponding SEM image, when 20 s exposure time was used; scale bar shown reads 100 nm) and b) corresponding temperature change measured in the dispersion, after exposing the dispersion with the ultrasonic probe.

the exposure time beyond 50 s does not seem to result in further breaking up of nanoparticles; it is hypothesised that shear energy provided by the ultrasonic probe was sufficient to de-aggregate but cannot sufficiently break nanoparticles that have fused together (i.e. agglomerates). A side effect of ultrasonication is the increase in temperature of the dispersion, due to the high shear energy that it provides [16]. Figure 2b) shows the change in temperature that occurred as a result of dispersion at various "exposure times" and the relationship between the two variables is approximately linear. Ideally, temperature changes in the dispersion should be minimised and so, a 20 s exposure time was chosen for the dispersion protocol as this gave ~5°C temperature increase (with corresponding particle size of 215 nm as shown in the DLS and the corresponding SEM image in Figure 2a). Results so far have shown the importance of having well controlled protocols for dispersions, as slight changes to procedures can potentially influence the particle size distribution of the resultant dispersion.

3.2 Limitations of characterisation tools

Figure 3a) and b) shows the effect of reducing CeO_2 nanoparticle concentration in DI water (from 500 mg/L to 0.001 mg/L) on DLS mean particle size and zeta-potential, respectively. Both plots show that values measured are similar above certain concentration. However, under extreme dilution conditions, data values shift significantly, yielding what is thought to be erroneous results. The erroneous data defines the limit of quantification for particle size and zeta-potential to be 0.1 and 50 mg/L, respectively. The results at extreme dilution is not surprising and explanations have been previously attributed: the inherent homodyne configuration of the optics a combination of: increase in signal contribution due to extraneous particles and inherent sensitivity of the detector, which defines the limit of quantification for particle size and zeta-potential, respectively [17]. Figure 3c) shows a series of SEM images of CeO_2 (dispersed in DI water and subsequently adsorbed on the surface of poly-L-lysine substrates), upon changing nanoparticle concentration within the dispersion. It is apparent that the particle size distribution changes dramatically when nanoparticle concentration in the dispersion is diluted from 500 mg/L to 10 mg/L. A much-reduced number of nanoparticles and a tendency for smaller particles adhering to the surface were observed under the extreme dilution conditions.

Unlike the DLS instrument, the limit of quantification is governed by the adsorption kinetics of the nanoparticles on to the substrates during the sample preparation step. At low nanoparticle concentrations, it is the diffusion rate of the particle that will dominate and this in turn explains why the smaller particles are preferentially adsorbed [12]. Overall, identifying the limit of quantification for an individual analytical procedure is important, so as to understand when data becomes unreliable. This is particularly of importance to nanoecotoxicological investigations as researchers in this field are often interested in making measurements under extreme dilute conditions, in the order of less than a few ng/L [10].

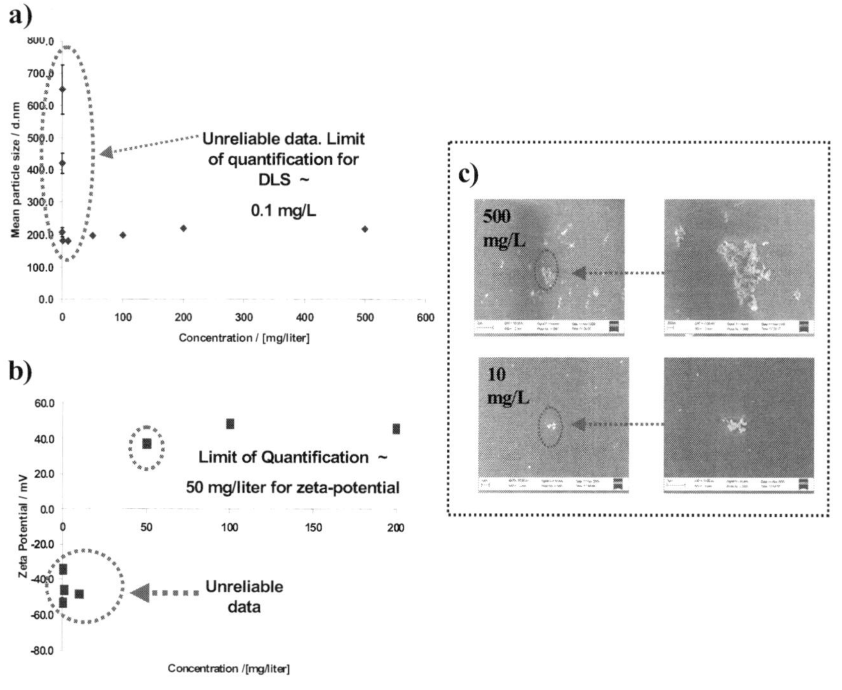

Figure 3: CeO_2 dispersed in DI water and the effect of varying nanoparticle concentration on: a) DLS mean particle size b) zeta-potential c) SEM of nanoparticles adsorbed on poly-L-lysine slides. Scale bars SEM: for 500 mg/L (reads 2 μm and 200 nm, for low and high magnification, respectively) and for 10 mg/L (reads 1 μm for both).

Figure 4 aims to identify where common laboratory techniques like DLS and SEM sit on the technology space map. Three important criteria have been identified as being essential: sensitivity (x-axis), selectivity (y-axis) and representativeness (z-axis). Ideally, an instrument should have a high degree of sensitivity (to single particle level), high selectivity (to measure in the presence of potentially interfering substances in the ecotox media [10] and high representativeness (such that the data is a representation of the entire population rather than a subset; this will subsequently contribute towards the accuracy and repeatability of the measurements); the ideal tool sits on vertex C. According to the technology road map, the DLS and SEM sit on vertex A and B, respectively. The DLS, belongs to a category of "population based methods"; they may not have the desired combination of high sensensitivity and selectivity but data are highly representative of the entire population. SEM on the other hand, belongs to a category of "single particle based methods", which are highly sensitive and selective but data obtained lacks "representativeness". Future research activities should therefore employ and subsequently validate a technique that belongs on vertex C on this space map.

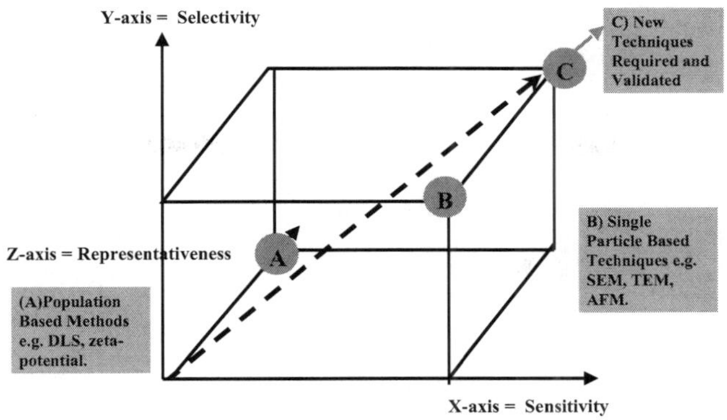

Figure 4: Technology space map: tools/techniques for nanoparticle characterisation in complex environmental media assessed against three identified criteria of: sensitivity (x-axis), selectivity (y-axis) and "representativeness" (z-axis).

3.3 Visual sedimentation

Figure 5 shows a typical result from the visual sedimentation experiment of when nanoparticles (either ZnO or CeO_2) are dispersed in one of the ecotox media (500 mg/L). After two days, the dispersion inside the bottle showed complete sedimentation, leaving a clear solution above the sediment. Interestingly, if nanoparticles were dispersed in DI water instead, a different result is apparent in that a cloudy suspension was still observed on Day 2. This suggested that the particles were more stable in DI water and subsequently a much slower sedimentation rate was observed. When salts were present in the ecotox media, this caused particle instability and resulted in aggregation/agglomeration [18]. The addition of salts in the ecotox liquid formulation employed in this study was sufficient to cause particle aggregation, resulting in a much faster sedimentation rate. If such sedimentation events occurred in ecotoxicological relevant media, then this raises the question as to whether there would be significant safety concerns. On Day 3, samples inside the bottles were analysed with an SEM; most particles present in the dispersion were shown to constitute largely of micron-sized particles, as expected. However, upon careful examination, some smaller clusters of both nanoparticles (particle diameter size of less than ~800 nm) were present in all four media, as shown in Figure 5; particle diameter size was estimated by assuming spherical shape cluster. From the SEM examination, this was the case for ZnO disperse in DI water, seawater and Daphnia liquid media; for the case of CeO_2, this was true upon dispersion in DI water and fish media. If the hypothesis that relates particle size with toxicity holds true, then it would be the presence of such nanosize clusters that should be of utmost concern to the aquatic environment.

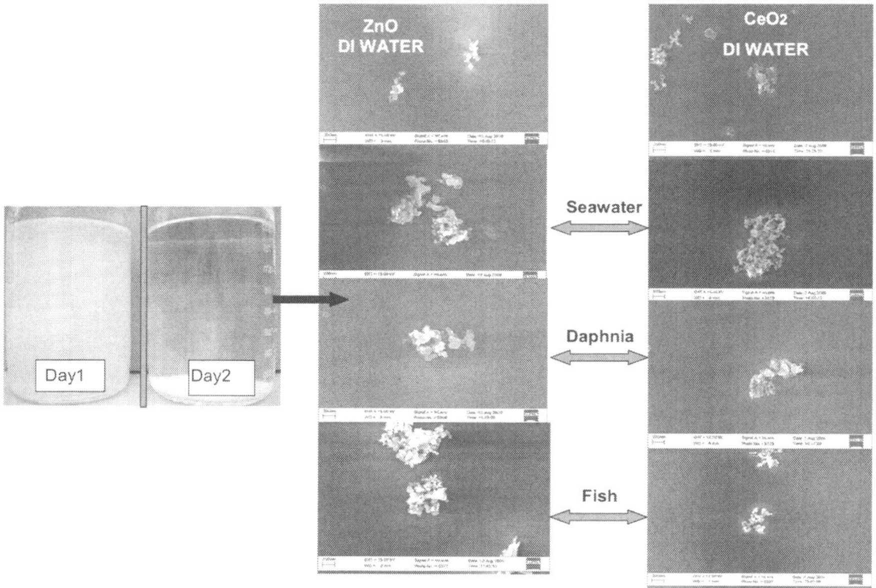

Figure 5: Visual sedimentation experiment of nanoparticles (typical of either ZnO or CeO_2 dispersed in an ecotox media (seawater or daphnia or fish media), during a two-day period. SEM images show the presence of nanosize clusters of nanoparticles present in the bottles for both ZnO and CeO_2 (obtained on third day); scale bar on all SEM reads 200 nm.

4 Conclusion

While most nanoparticles were shown to aggregate out of solution when immersed in ecotoxicological media, some nanosized clusters were still present. If this was to occur in a real environmental setting, then there is potential for aquatic organisms to ingest such small particles. Through time, these small particles can accumulate and it is the accumulation of dose that can subsequently formulate a problem. In terms of protocol developments in ecotoxicological investigations, there is a need to:

a) have well controlled, agreed protocols on dispersion and characterisation
b) identify, develop and validate suitable tools / technology that can offer a combination of high sensitivity, selectivity and "representativeness"

Acknowledgements

This work is financed under Defra PROSPEcT LINK project. The authors would like to thank Mr. Jordan Tompkins for initial handling and distribution of the nanoparticles; Neil Harrison and Alex Shard for helpful comments and encouragement.

References

[1] Maynard, A.D., Pui, D.Y.H., *Nanotechnology and occupational health: New technologies - new challenges.* Journal of Nanoparticle Research, 2007. **9**(1): p. 1-3.

[2] Oberdorster, G., Stone, V., Donaldson, K., *Toxicology of nanoparticles: A historical perspective.* Nanotoxicology, 2007. **1**(1): p. 2-25.

[3] Nel, A., Xia, T., Madler, L., Li, N., *Toxic potential of materials at the nanolevel.* Science, 2006. **311**(5761): p. 622-627.

[4] Durnev, A.D., *Toxicology of nanoparticles.* Bulletin of Experimental Biology and Medicine, 2008. **145**(1): p. 72-74.

[5] Jaurand, M.C.F., Renier, A., Daubriac, J., *Mesothelioma: Do asbestos and carbon nanotubes pose the same health risk?* Particle and Fibre Toxicology, 2009. **6**: p. 14.

[6] Poland, C.A., Duffin, R., Kinloch, I., Maynard, A., Wallace, W. A. H., Seaton, A., Stone, V., Brown, S., MacNee, W., Donaldson, K., *Carbon nanotubes introduced into the abdominal cavity of mice show asbestos-like pathogenicity in a pilot study.* Nature Nanotechnology, 2008. **3**(7): p. 423-428.

[7] Koyama, S., Endo, M., Kim, Y. A., Hayashi, T., Yanagisawa, T., Osaka, K., Koyama, H., Haniu, H., Kuroiwa, N., *Role of systemic T-cells and histopathological aspects after subcutaneous implantation of various carbon nanotubes in mice.* Carbon, 2006. **44**(6): p. 1079-1092.

[8] Tantra, R., Tompkins, J. and Quincey P., *Characterisation of the de-agglomeration effects of bovine serum albumin on nanoparticles in aqueous suspension.* 2010. **75**: p. 275-281.

[9] Necula, B.S., Apachitei, I., Fratila-Apachitei, L., Teodosiu, C., Duszczyk, J., *Stability of nano-/microsized particles in deionized water and electroless nickel solutions.* Journal of Colloid and Interface Science, 2007. **314**(2): p. 514-522.

[10] Simonet, B.M., Valcarcel, M., *Monitoring nanoparticles in the environment.* . Anal. Bioanal. Chem., 2009. **393**: p. 17-21.

[11] Lundqvist, M., Stigler, J., Elia, G., Lynch, I., Cedervall, T., Dawson, K. A., *Nanoparticle size and surface properties determine the protein corona with possible implications for biological impacts.* Proceedings of the National Academy of Sciences of the United States of America, 2008. **105**(38): p. 14265-14270.

[12] Kissa, E.E., *Dispersions: characterization, testing, and measurement* 1999: CRC Press.

[13] *ISO 14887:2000 Sample preparation -- Dispersing procedures for powders in liquids.*

[14] Rhein, L.D., Schlossman M., O'Lenick A. and Somasundaran P. (Editors) and E. 3, *Surfactants in personal care products and decorative cosmetics.* 2006: Publisher CRC Press.

[15] Jailani, S., Franks, G. V., Healy, T. W., *Zeta-potential of nanoparticle suspensions: Effect of electrolyte concentration, particle size, and volume*

fraction. Journal of the American Ceramic Society, 2008. **91**(4): p. 1141-1147.
[16] Kanegsberg, B., Kanegsberg, E., *Handbook for critical cleaning* 2001: CRC Press.
[17] Tantra, R., Schulze P. and Quincey, P., *Effect of Nanoparticle Concentration on Zeta-Potential Meausurement Results and Reproducibility.* Particuology, *in press*.
[18] Murdock, R.C., Braydich-Stolle, L., Schrand, A. M., Schlager, J. J., Hussain, S. M., *Characterization of nanomaterial dispersion in solution prior to In vitro exposure using dynamic light scattering technique.* Toxicological Sciences, 2008. **101**(2): p. 239-253.

High-throughput analysis of multiple stress pathways using GFP reporters in *C. elegans*

D. de Pomerai[1], C. Anbalagan[1], I. Lafayette[1], D. Rajagopalan[1], M. Loose[1], M. Haque[2] & J. King[2]
[1]*School of Biology, University of Nottingham, UK*
[2]*School of Mathematical Sciences, University of Nottingham, UK*

Abstract

Stress-responsive genes belonging to multiple defensive pathways in the nematode *C. elegans* are cross-regulated by kinase signalling (AKT-1/-2, p38 MAPK) and transcription factors (DAF-16, SKN-1). This cross-talk between stress pathways implies that they are best regarded as a stress-response network (SRN), whose behaviour as a whole should be amenable to mathematical modelling. We have used GFP reporter strains to provide a rapid readout of expression levels for 24 genes, representing principal outputs and transcription factors in the heat-shock, metal-binding, oxidative stress, phase I & phase II detoxification, and genotoxic stress pathways. Acute toxicity data (up to ~24 h) has been generated for selected metal (presented here) and pesticide toxicants across a wide range of doses, and common response patterns identified. Mathematical modelling of these response data, informed by an understanding of the underlying genetic circuitry, should allow our model to predict the likely toxicity of pollutant mixtures. Future work will test the accuracy of such predictions, leading to an iterative process of model refinement.

Keywords: metal toxicity, caenorhabditis elegans, mathematical modelling, mixture toxicity, high-throughput GFP reporter assays, stress-response network.

1 Introduction

Both chemical and physical (e.g. heat) stressors evoke defensive responses in living organisms – including DNA repair to counteract genotoxic DNA damage, heat-shock protein (hsp) expression to counter proteotoxic (e.g. thermal) damage to proteins, metal-binding proteins to sequester heavy metals, phase I and phase

WIT Transactions on Ecology and the Environment, Vol 132, © 2010 WIT Press
www.witpress.com, ISSN 1743-3541 (on-line)
doi:10.2495/ETOX100171

II detoxification enzymes to metabolise complex organic compounds, and oxidative stress enzymes to deal with reactive oxygen species (ROS). Several of these pathways can be activated by a single agent: cadmium (Cd) is genotoxic, binds to reduced glutathione thus increasing ROS levels, and is a potent inducer of hsp expression. Paradoxically, cadmium can also induce the expression of multiple cytochrome P450 (*cyp*) genes, as revealed by gene-array studies in the nematode *C. elegans* [1]; we and others have demonstrated that for the *cyp*-34A9 gene at least, Cd regulation is probably mediated by the DAF-16 transcription factor regulating longevity [2, 3]. Indeed, DAF-16 acts as a positive regulator for many other stress-response genes, including *sod*-3 (mitochondrial superoxide dismutase) and *mtl*-1 (metal-binding metallothionein) as well as *hsp*-16.1 and *hsp*-16.2 (encoding small hsp's) [3]. Thus functionally distinct stress-response pathways are clearly cross-regulated by core transcription factors in *C. elegans*.

C. elegans is a useful genetic model in which to examine this stress-response network (SRN) as a whole, thanks to its ease of culture, short life cycle, excellent genetics, complete genome sequence and numerous post-genomic tools including feeding RNAi screens to explore gene functions. In the context of environmental toxicology, it also has certain disadvantages – being relatively resistant to most stressors thanks to an impressive armoury of defensive pathways, including over 80 *cyp* phase I and 40 *gst* (glutathione S-transferase) phase II genes. One obvious way to explore genome-wide transcriptional responses to stressors is to use gene arrays, but cost considerations preclude this for routine screening of multiple toxicants over a range of doses, let alone applying this methodology to mixtures. We have adopted an alternative approach employing transcriptional GFP fusion genes (with an EGFP coding sequence driven by ~3 kb of upstream promoter). Integrated transgenic strains carrying multiple copies of such GFP reporter genes have been provided for us by the Baillie Genome GFP Project or obtained from CGC and other sources. GFP fluorescence levels provide a high-throughput quantitative measure of transgene expression in real time, allowing several time points as well as multiple stress genes, toxicants and test doses to be compared. We aim to develop a predictive mathematical model describing SRN behaviour.

This paper reports GFP expression patterns for 24 test genes in response to 7 metals (Zn^{2+}, Cd^{2+}, Hg^{2+}, Cu^{2+} , Fe^{3+}, Cr^{6+}, As^{3+}), as well as preliminary data on metal mixture toxicity. The genes chosen include many of those involved in combating oxidative stress (*sod*-1, *sod*-3, *sod*-4, *ctl*-2; & 2 uncurated glutathione peroxidise genes – T09A12.2 designated GPA and C11E4.1 designated GPB), 2 metallothionein genes (*mtl*-1 and *mtl*-2) [4] and several heat-shock genes (*hsp*-3, *hsp*-3, *hsp*-60, *hsp*-16.1, *hsp*-16.2; & the uncurated C12C8.1 gene encoding a major cytosolic HSP70 designated cHSP70). Genes encoding key transcription factors involved in regulating these pathways are also included:- *skn*-1 (affecting oxidative stress genes) [5], *hsf*-1 (regulating heat-inducible *hsp* genes), *daf*-16 [3, 6], and *elt*-2 (controlling adult intestinal genes including both *mtl* genes) [7], as well as the p53 orthologue *cep*-1 involved in responding to genotoxic stress [8]. Among the many phase 1 and phase II genes, only a few candidates already known to be stress-inducible have been included:- these are *cyp*-29A2, *cyp*-34A9 and *cyp*-35A2 [9], as well as *gst*-1 and *gst*-4 [10, 11].

2 Materials and methods

2.1 Materials

PC161 (*hsp*-16.1::GFP:*lacZ*) was developed in-house [12]; grateful thanks are extended to Cynthia Kenyon for CF1553 (*sod*-3::GFP), Chris Link for CL2050 (*hsp*-16.2::GFP), Joel Rothman for JR2474 (*cep*-1::GFP), Ralph Menzel for a strain with pPD97 87-35A2prIII-GFP (*cyp*-35A2::GFP) and the *Caenorhabditis* Genetics Center (funded by the NIH National Center for Research Resources) for TJ356 (*daf*-16::GFP). Other strains were supplied as integrated promoter::GFP fusions by the Baillie Genome GFP Project (Simon Fraser University, Burnaby, Vancouver, Canada), each containing about 3 kb of upstream promoter sequence (apart from BC20306 which contains only the 250 bp of promoter separating the *cyp*-34A9 gene from its upstream neighbour *cyp*-34A10) [2, 13]. These are:- BC17553 (T09A12.2 glutathione peroxidase designated GPA::GFP), BC20303 (*hsp*-6::GFP), BC20305 (C11E4.1 glutathione peroxidise designated GPB::GFP), BC20306 (*cyp*-34A9::GFP), BC20308 (*hsp*-3::GFP), BC20309 (*mtl*-1::GFP), BC20314 (*elt*-2::GFP), BC20316 (*gst*-1::GFP), BC20329 (*skn*-1::GFP), BC20330 (*gst*-4::GFP), BC20333 (*sod*-4::GFP), BC20334 (*cyp*-29A2::GFP), BC20336 (*ctl*-2::GFP), BC20337 (*hsf*-1::GFP), BC20342 (*mtl*-2::GFP), BC20343 (*hsp*-60::GFP), BC20349 (C12C8.1 *hsp*-70 designated cHSP70::GFP) and BC20350 (*sod*-1::GFP). All patterns of induced and uninduced GFP expression have been verified against expression data (where known) for the corresponding gene on Wormbase. All chemicals were from Sigma Ltd and plasticware from Nunc Ltd.

2.2 Methods

All GFP reporter strains were grown at 15^0C on NGM agar plates with a lawn of P90C *E. coli* as food [14]; worms were harvested after 3-4 days of culture, and equal aliquots of ~500 worms (all stages) were exposed to increasing sublethal doses of metal ions diluted in K medium (53 mM NaCl, 32 mM KCl) [15]. The dose ranges tested were:- 0.32-200 ppm Zn^{2+}, 0.022-22 ppm Cd^{2+}, 0.016-10 ppm Hg^{2+}, 0.06-15 ppm Cu^{2+}, 0.01-100 ppm Fe^{3+}, 0.064-40 ppm Cr^{6+} (as $Cr_2O_7^{2-}$) and 0.01-100 ppm As^{3+} (as AsO_2^-) [15]; in all cases, a set of zero controls (no metal) was also included. After 4-6 (early), 8-12 (intermediate) or 24-30 (late) hours of exposure, all worms were transferred into black non-fluorescent 96-well plates and the levels of GFP fluorescence measured in a Perkin-Elmer Victor 1420 plate fluorometer. Full details of this methodology will be published separately. Four independent assays were conducted for each test condition, and the means ± SEMs compared using one-way ANOVA and Dunnett's multiple comparisons test. Typical GFP data are shown in Figure 1 for three genes responding to Hg^{2+}.

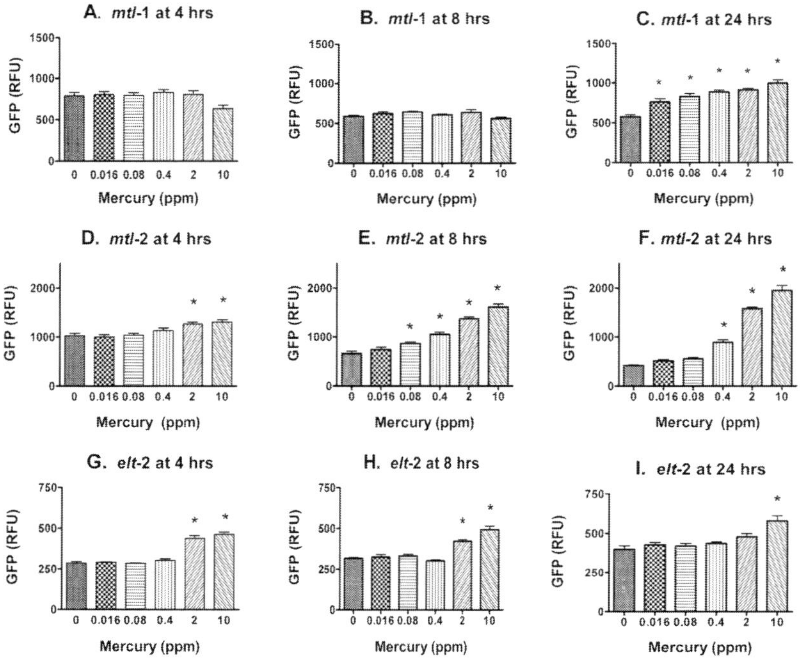

Figure 1: Expression of metallothionein-related GFP reporters in response to Hg. Reporter GFP strains transgenic for *mtl*-1 (A-C), *mtl*-2 (D-F) or *elt*-2 (G-I) were exposed to the indicated concentrations of $HgCl_2$ (5-fold dilutions from 10 ppm down to 16 ppb, plus zero controls) for 4 hr (A, D, G), 8 hr (B, E, H) or 24 hr (C, F, I), and GFP expression was measured (in relative fluorescence units, RFU) as described in methods. After applying one-way ANOVA and Dunnett's *post hoc* multiple comparisons test to the data within each figure, expression levels that differ significantly from zero controls at the same time point are indicated by asterisks ($p < 0.05$; $n = 4$ in all cases).

3 Results

By its very nature, a high-throughput study of multiple genes and toxicants generates large volumes of data. The first stage in data compilation is exemplified by Figure 1, which shows the responses to Hg^{2+} of 3 functionally related genes (the 2 *C. elegans* metallothionein genes, *mtl*-1 and *mtl*-2, plus their candidate transcriptional activator *elt*-2), comparing a range of 5-fold doses (0 plus 16 ppb up to 10 ppm) at 3 different time points (4, 8 and 24 hours). GFP expression levels that differ significantly from zero-Hg controls are indicated by asterisks ($p < 0.05$ using Dunnett's test). Broadly speaking, *mtl*-1 shows less clear-cut induction than *mtl*-2, in agreement with previous findings [4]; the *mtl*-1 gene is expressed constitutively in the posterior pharynx, and is inducible in the intestine of larvae, but less so in adults [4]. Since assays were conducted on mixed stage cultures, it is possible that the stronger induction of this gene seen at 24 hrs (significant at all test doses) reflects *mtl*-1 induction in a new generation of young larvae. Notably, *mtl*-2 expression is already induced at 4 hrs at the highest test doses, and shows clear dose-dependent induction at the later time points. By contrast, *elt*-2 expression is only activated at the two highest concentrations of Hg^{2+}, and even this response fades out somewhat by 24 hrs.

It would clearly be impractical to extend this presentation to 6 other metals and to all 24 test genes included in this study. A first-level simplification is shown in Table 1, which summarises the responses to Hg^{2+} of a larger group of 7 heat-shock related genes (*hsp*-3, *hsp*-6, *hsp*-60, *hsp*-16.1, *hsp*-16.2, the uncurated C12C8.1 gene encoding an inducible cHSP70, and their key transcription factor gene, *hsf*-1). For simplicity, the early response data (at 4 hrs) have been omitted, as only *hsp*-16.1 showed any response at this time point. Rather than presenting measured GFP data in both zero controls and at each test dose, Table 1 presents simple ratios (mean GFP level at test dose divided by mean GFP level in zero control at that time point) indicating relative induction or repression of transgene expression. In accordance with the common practice for gene array data, Table 1 highlights significant ($p < 0.05$) expression changes of > 50% (up or down), and also underlines all changes > 2-fold. A greater diversity of expression responses is apparent here, with some genes (e.g. *hsp*-16.1) showing greater sensitivity to Hg^{2+} than others, while *hsp*-60 is virtually unaffected. As for Figure 1, the core transcription factor (here *hsf*-1) is much less sensitive than many of its target genes; however, it too is activated at the highest test dose at 8 hrs, suggesting that additional transcription factor expression can be deployed in an emergency.

A final level of simplification is shown in Table 2 for the complete set of 7 metals and 24 GFP transgenes. The entries in each cell record only the maximal effect seen (at any dose or time point) for that toxicant on that test gene. Broadly speaking, + or – symbols indicate a statistically significant ($p < 0.05$) rounded change of between 1.5- and 2-fold (up or down), while ++ indicates a 2- to 3-fold and +++ a > 3-fold up-regulation of expression (no cases of > 2-fold down-regulation were seen in this data set); the maximal induction seen was about 4.5-fold. Some general trends apparent within this large data set are discussed below. Whilst much important detail has been lost, a more complete overview emerges.

One of the major aims of this collaborative study is to develop a predictive mathematical model describing the behaviour of the SRN as a whole. As a start, the mathematicians involved (MH and JK) have generated initial models of two regulatory sub-networks – one linking the ELT-2 transcription factor to the 2 *mtl* genes, and the other linking HSF-1 to inducible heat-shock genes; only the former is considered here. A decade ago, Moilanen *et al* [16] suggested that both *mtl* genes are positively regulated by ELT-2, which has recently been implicated as the major transcription factor maintaining intestine-specific patterns of gene expression in the adult [7]. We have used RNAi to show that ELT-2 is required

Table 1: Expression of heat-shock related GFP reporter transgenes following exposure to mercuric chloride. This table presents a first-level simplification of the processed data (Figure 1), and shows the responses to Hg^{2+} of 6 *hsp* genes (N.B. C12C8.1 is uncurated, but encodes a cytoplasmic HSP70, designated cHSP70) plus the transcription factor *hsf*-1. The methodology was the same as in Figure 1, but results are expressed as ratios between GFP expression at the test dose of Hg^{2+} and in corresponding zero controls (n = 4). Statistically significant differences in expression (Dunnett's test) that exceed 1.5-fold are highlighted in grey, and those that exceed 2-fold are further emphasised by underlining. Early responses at 4 hours have been omitted, since only *hsp*-16.1 shows any response in this range (expression ratio 1.58 at 10 ppm Hg^{2+}). One marginal case (almost 1.5-fold up-regulation at 24 hr for cHSP70) is shown in bold italics for emphasis.

	Intermediate GFP expression response at 8 hr (ratio, as compared to zero control).							Late GFP expression response at 24 hr (ratio, as compared to zero control).					
Hg dose Gene	0 ppm	0.016 ppm	0.08 ppm	0.4 ppm	2 ppm	10 ppm		0 ppm	0.016 ppm	0.08 ppm	0.4 ppm	2 ppm	10 ppm
hsp-16.1	1.00	1.20	1.28	1.33	1.82	2.90		1.00	1.22	1.24	1.33	1.63	2.64
hsp-16.2	1.00	1.25	1.31	1.37	1.44	1.85		1.00	1.26	1.52	1.34	1.45	1.92
hsp-3	1.00	1.25	1.30	1.45	1.74	1.57		1.00	1.12	1.20	1.22	1.46	1.45
hsp-6	1.00	1.26	1.38	1.41	1.72	2.09		1.00	1.06	1.15	1.17	1.36	1.72
hsp-60	1.00	1.10	0.99	1.04	1.10	1.42		1.00	1.04	0.94	0.90	1.09	1.47
cHSP70 C12C8.1	1.00	1.19	1.19	1.27	1.35	1.28		1.00	1.16	1.28	1.27	1.35	***1.49***
hsf-1	1.00	1.22	1.28	1.26	1.34	1.62		1.00	1.16	1.17	1.14	1.15	1.29

Table 2: Summary of responses for 7 metals in all 24 GFP transgenic strains. These table entries summarise the maximum responses seen across the sublethal doses tested, without distinguishing whether these are dose-dependent or seen only at the maximum dose, nor whether they were observed early (rarely), or at intermediate or late time points. In all cases, no entry in a cell signifies a test:control ratio of 0.67 to 1.45 ($<$ 1.5-fold change); + indicates a ratio of 1.45 to 1.94 (1.5-2 fold up); ++ indicates a ratio of 1.95 to 2.94 (2-3 fold up); +++ indicates a ratio $>$ 2.95 (3-fold up); and - indicates a ratio of 0.66 to 0.5 (1.5-2 fold down). All such expression changes are statistically significant ($p < 0.05$; $n = 4$). Ratios have in all cases been rounded to the first decimal place.

Metal → / Gene ↓	Zn^{2+}	Cd^{2+}	Hg^{2+}	Cu^{2+}	Fe^{3+}	Cr^{6+}	As^{3+}
mtl-1	++	+	+	++			+
mtl-2	++	+++	+++	+	++		+
elt-2	++		+	++		++	
hsp-16.1	++	+++	++	+	+++	+	++
hsp-16.2	++	++	+	+++	+++	+	+++
hsp-3	+	+++	+	+++	+		
hsp-6	+	++	++	++	+	+	++
hsp-60	++	++	+	+++		++	++
cHSP70 C12C8.1	+	+	+	++		++	+
hsf-1	+	+	+	+	-		
sod-1	+		+	+		++	
sod-3	-		+	++			
sod-4	++	++	++		+++	+	+
ctl-2	+		+		-		
GPA T09A12.2	++		+	++	++		
GPB C11E4.1	++	+	+	++	++		++
skn-1	+		+				
gst-1	+++	+	+	++	-	+	++
gst-4	+	++	+	++		++	+
cyp-29A2	++	++	+	++			+
cyp-34A9	++	++	+	+	+		++
cyp-35A2	+++	+	++	++	+	+	++
cep-1	++		+	+		+	+
daf-16	+				++		+

for Cd or Hg induction of *mtl*-2::GFP expression, though the effect on *mtl*-1::GFP is less clear-cut (CA and DdeP, unpublished data). Because *elt*-2 and most intestine-specific genes are expressed constitutively, the low levels of intestinal *mtl* gene expression in the absence of metal induction additionally require an unknown metal-sensitive repressor [16] – still elusive despite a decade of further research. However, since ectopic expression of ELT-2 activates *mtl* expression outside the intestine in the absence of metals, this key feature of the model seem to be vindicated. This genetic circuit forms the basis of our mathematical model, which will be published separately (MH, DdeP and JK, in preparation). The broad quantitative patterns of *mtl* gene induction (Figure 1; top 2 lines of Table 2) are consistent with metal binding (both to MTL proteins and to the unknown repressor) being very much more rapid than the inducibility of *mtl* gene expression promoted by ELT-2. Assuming that the system does not reach equilibrium within the time course tested (up to 24 hours), then our model predicts that chemically similar divalent metal ions, such as Zn, Cd and Hg (all of which bind to MTLs) should show additive effects on *mtl*-2::GFP transgene expression, as confirmed for sub-maximal doses of Cd^{2+} and Hg^{2+} in Figure 2A. However, *mtl*-2 is also inducible by trivalent Fe^{3+} at high concentrations (~100 ppm), although lower doses have little effect. Surprisingly, mixtures of Fe^{3+} with any of the divalent group IIB metals show interference, i.e. reduced *mtl*-2::GFP expression compared with the divalent ion on its own (shown for Fe^{3+} and Hg^{2+} in Figure 2B). This can be accommodated by the mathematical model if Fe^{3+} interferes with divalent-ion binding to MTLs and/or to the unknown repressor.

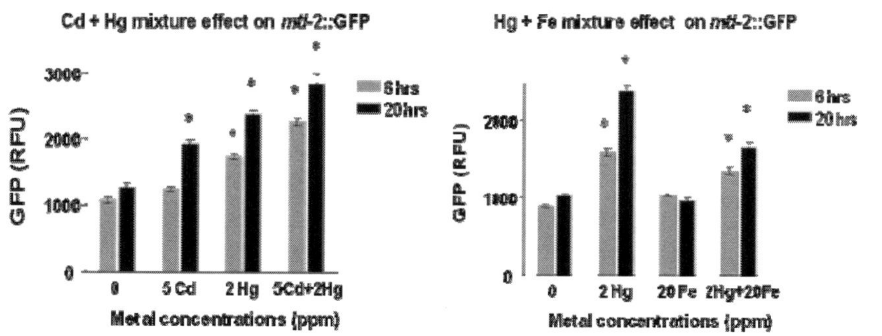

A. B.

Figure 2: Selected examples of metal-mixture toxicity effects on *mtl*-2::GFP. Part A: submaximal concentrations of Cd^{2+} (5 ppm) and Hg^{2+} (2 ppm) tested both singly and together (plus 0 control), and showing a mildly additive response. Part B: submaximal concentrations of Hg^{2+} (2 ppm) and Fe^{3+} (20 ppm) tested both singly and together (plus 0 control), showing Fe^{3+} inhibition of Hg^{2+} induction in this mixture. Significant changes ($p < 0.01$) relative to zero controls are indicated by asterisks.

4 Discussion

High-throughput monitoring of GFP reporter transgene expression is obviously much cheaper than using gene array approaches, though only a targeted selection of genes can be covered. This limitation is particularly apparent in the case of large gene families such as the *cyp* and *gst* groups. For other stress-responsive pathways, our coverage is around 50% and could easily be extended further. Gene expression changes outside the SRN are not covered, which may exclude important response genes (e.g. the *cdr* or *aip* genes responding to Cd or As). The use of GFP reporters as transcriptional fusion genes also imposes limitations, since dynamic changes in gene expression will not be reflected in the levels of relatively stable GFP; translational fusions and destabilised GFP variants could be used to circumvent this problem in future. Furthermore, some time (of the order of 1-2 hrs) is required for GFP to be translated, folded and auto-oxidised before any fluorescent signal can be detected; thus a lack of response at 4 hrs need not imply that the gene under investigation still remains to be induced.

Even acknowledging these limitations, Table 2 shows how this approach can provide a more global view of (in this case) patterns of stress-gene response to a series of metals [cf. LC50 data in 15]. Broadly similar gene expression patterns are induced by the 3 group IIB metals (Zn^{2+}, Cd^{2+}, Hg^{2+}) and by Cu^{2+}, whilst more divergent patterns are apparent for Fe^{3+}, Cr^{6+} and the metalloid As^{3+}. As expected, both *mtl* genes are inducible by divalent metals – *mtl*-2 more strongly than *mtl*-1 [4, 16]. Responses to As^{3+} and to Fe^{3+} (*mtl*-2 only) may reflect interference with the handling of divalent metals, as implied also by mathematical modelling and mixture effects shown in Figure 2 (see end of Results). Among the *hsp* group, the two *hsp*-16 genes show the strongest responses (apart from Cr^{6+}), while the *hsp*-3 gene (encoding an ER-specific HSP70) is most strongly induced by Cd^{2+} and Cu^{2+}. The oxidative stress genes are mostly rather weakly induced, apart from the extracellular superoxide dismutase encoded by the *sod*-4 gene. Of the two glutathione peroxidase genes tested, C11E4.1 (GPB) expressed in intestine shows a somewhat greater range of metal sensitivity than is the case for T09A12.2 (GPA) expressed in body wall. Perhaps most surprising is the strong induction of all three *cyp* genes by most of the metals tested (less so by Cr^{6+} and Fe^{3+}), perhaps reflecting metal-induced changes in general metabolism, since metals themselves cannot be metabolised by cytochrome P450-mediated processes such as epoxidation. This feature has also been noted by others, since many *cyp* genes are prominently up-regulated by Cd^{2+} in gene array studies [1]. In general, the transcription factor (TF) genes studied (*elt*-2, *hsf*-1, *skn*-1, *cep*-1 and *daf*-16) show fewer and weaker responses to metal stressors, though *cep*-1 is strongly induced by Zn^{2+} and *daf*-16 by Fe^{3+}. These TFs are expressed widely in *C. elegans*; thus increased expression of their genes may represent a last-ditch attempt to up-regulate appropriate target genes. Although many previous studies have compared the responses of single SRN output genes to multiple stressors (e.g. [4, 10–12, 14, 16]), this is the first report to make such comparisons for multiple output genes in different stress pathways.

In summary, this GFP-reporter approach provides a broad overview of stress-gene expression patterns that would otherwise need many expensive gene arrays. These quantitative data can be incorporated into mathematical modelling of the SRN and of its component sub-networks, based on the known genetic circuitry of these systems. One advantage of such modelling is that it can generate rational predictions about the likely toxicity of mixtures; thus the mixture data shown in Figure 2 can be accommodated by a mathematical model of the *mtl* sub-network. Future work will extend this modelling, first to other regulatory sub-networks (e.g. the oxidative stress genes controlled by SKN-1) [5], and then to the entire SRN. Model-based predictions of likely mixture toxicity can be readily tested in the laboratory – and if found to be inaccurate, the model parameters can be adjusted accordingly. Through this iterative process of model refinement, model predictions for simple toxicant mixtures should become increasingly accurate over time. This represents a rational approach to a key problem in environmental toxicology – namely the fact that pollutants rarely come singly, and extrapolating the toxicity of mixtures from that of single constituents has rarely proved simple.

Acknowledgements

The authors gratefully acknowledge the contributions of David Baillie and Bob Johnsen (Simon Fraser University, Vancouver, Canada) for providing transgenic BC strains, and of UK-IERI (Major Award MA-05) for their financial support.

References

[1] Cui, Y., McBride, S.J., Boyd, W.A., Alper, S. & Freedman, J.H., Toxicogenomic analysis of *Caenorhabditis elegans* reveals novel genes and pathways involved in the resistance to cadmium toxicity, *Genome Biology* **8**, R122, 2007.

[2] de Pomerai, D., Madhamshettiwar, P., Anbalagan, C., Loose, M., Haque, M., King, J., Kar Chowdhuri, D., Sinha, P., Johnsen, R. & Baillie, D., The stress-response network in animals: development of a predictive mathematical model, *Open Toxicology Journal* **2**, pp. 71-76, 2008.

[3] Murphy, C. McCarroll, S.A., Bargmann, C.I., Fraser, A., Kamath, R.S., Ahringer, J., Li, H. & Kenyon, C., Genes that act downstream of DAF-1 to influence the lifespan of *C. elegans*, *Nature* **424**, pp. 277-84, 2003.

[4] Swain, S.C., Keusekotten, K., Baumeister, R. & Sturzenbaum, S.R., *C. elegans* metallothioneins: new insights into the phenotypic effects of cadmium toxicosis, *Journal of Molecular Biology* **341**, pp. 951-959, 2004.

[5] An, J.-H., & Blackwell, T., SKN-1 links *C. elegans* mesendodermal specification to a conserved oxidative stress response, *Genes and Development* **17**, pp. 1882-1893, 2003.

[6] McElwee, J.J., Schuster, E., Blanc, E., Thomas, J.H. & Gems, D., Shared transcriptional signature in *Caenorhabditis elegans* dauer larvae and long-lived *daf*-2 mutants implicates detoxification systems in longevity assurance, *Journal of Biological Chemistry* **279**, pp. 44533-44543, 2004.

[7] McGhee, J.D., Fukushige, T., Krause, M.W., Minnema, S.E., Goszczynski, B., Gaudet, J., Kohara, Y., Bossinger, O., Zhao, Y., Khattra, J., Hirst, M., Jones, S.J.M., Marra, M.A., Ruzanov, P., Warner, A., Zapf, R., Moerman, D.G. & Kalb, J.M., ELT-2 is the predominant transcription factor controlling differentiation and function of the *C. elegans* intestine, from embryo to adult, *Developmental Biology* **327**, pp. 551-565, 2009.

[8] Derry, W.B., Putzke, A. & Rothman, J., *Caenorhabditis elegans* p53: role in apoptosis, meiosis, and stress resistance, *Science* **294**, pp 591-595, 2001.

[9] Menzel, R., Bogaert, T. & Achazi, R., A systematic gene expression screen of *Caenorhabditis elegans* cytochrome P450 genes reveals CYP35 as strongly xenobiotic inducible, *Archives of Biochemistry and Biophysics* **395**, pp. 158-168, 2001.

[10] Leiers, B., Kampkotter, A., Grevelding, C.G., Link, C.D., Johnson, T.E. & Henkle-Duhrsen, K., A stress-response glutathione S-transferase confers resistance to oxidative stress in *Caenorhabditis elegans*, *Free Radical Biology and Medicine* **34**, pp. 1405-1415, 2003.

[11] Hasegawa, K., Miwa, S., Isomura, K., Tsutsumiuchi, K., Taniguchi, H. & Miwa, J., Acrylamide-responsive genes in the nematode *Caenorhabditis elegans*, *Toxicological Sciences* **101**, pp. 215-225, 2008.

[12] David, H.E., Dawe, A.S., de Pomerai, D.I., Jones, D., Candido, E.P.M., & Daniells, C., Construction and evaluation of a transgenic *hsp16*-GFP-*lacZ* *Caenorhabditis elegans* strain for environmental monitoring, *Environmental Toxicology and Chemistry* **22**, pp. 111-118, 2003.

[13] Hunt-Newbury, R., Viveiros, R., Johnsenn, R., Mah, A., Anastis, D., Fang, L., Halfnight, E., Lee, D., Lin, J., Lorch, A., McKay, S., Okada, H.M., Pan, J., Schultz, A.K., Tu, D., Wong, K., Zhao, Z., Alexeyenko, A., Burglin, T., Sonnhammer, E., Schnabel, R., Jones, S.J., Marra, M.A., Baillie, D.L. & Moerman, D.G., High throughput in vivo analysis of gene expression in *C. elegans*, *PLoS Biology* **5**, e237, 2007.

[14] Dennis, J.L., Mutwakil, M.H.A.Z., Lowe, K.C. & de Pomerai, D.I., Effects of metal ions in combination with a non-ionic detergent on stress responses in a transgenic nematode, *Aquatic Toxicology* **40**, pp. 37-50, 1999.

[15] Williams, P.L. & Dusenbery, D.B., Aquatic toxicity testing using the nematode *Caenorhabditis elegans*, *Environmental Toxicology and Chemistry* **9**, pp. 1285-1290, 1990.

[16] Moilanen, L.H., Fukushige, T. & Freedman, J.H., Regulation of metallothionein gene transcription: identification of upstream regulatory elements and transcription factors responsible for cell-specific expression of the metallothionein genes from *Caenorhabditis elegans*, *Journal of Biological Chemistry* **274**, pp. 29655-29665, 1999.

Author Index

Biological Monitoring

Theory and Applications

Edited by: ***M.E. CONTI****, University of Rome 'La Sapienza', Italy*

Provides the reader with a basic understanding of the use of bioindicators both in assessing environmental quality and as a means of support in environmental impact assessment (EIA) procedures. The book primarily deals with the applicability of these studies with regard to research results concerning the basal quality of ecosystems and from an industrial perspective, where evaluations prior to the construction of major projects (often industrial plants) are extremely important.

Environmental pollution and related human health concerns have now reached critical levels in many areas of the world. International programs for researching, monitoring and preventing the causes of these phenomena are ongoing in many countries.

There is an imperative call for reliable and cost-effective information on the basal pollution levels both for areas already involved in intense industrial activities, and for sites with industrial development potential.

Biomonitoring methods can be used as unfailing tools for the control of contaminated areas, as well as in environmental prevention studies. Human biomonitoring is now widely recognized as a tool for human exposure assessment, providing suitable and useful indications of the 'internal dose' of chemical agents.

Bioindicators, biomonitors, and biomarkers are all well-known terms among environmental scientists, although their meanings are sometimes misrepresented. Therefore, a better and full comprehension of the role of biological monitoring, and its procedures for evaluating polluting impacts on environment and health, is needed. This book gives an overview of the state of the art of relevant aspects of biological monitoring for the evaluation of ecosystem quality and human health.

Series: The Sustainable World, Vol 17
ISBN: 978-1-84564-002-6 eISBN: 978-1-84564-302-7
2008 256pp £84.00